NUREG-1569

Standard Review Plan for In Situ Leach Uranium Extraction License Applications

I0488578

Final Report

Date Completed: June 2003
Date Published: June 2003

Prepared by
J. Lusher

J. Lusher, NRC Project Manager

Division of Fuel Cycle Safety and Safeguards
Office of Nuclear Material Safety and Safeguards
U.S. Nuclear Regulatory Commission
Washington, DC 20555-0001

NOTICE OF AVAILABILITY OF A STANDARD REVIEW PLAN (NUREG–1569) FOR STAFF REVIEWS FOR *IN SITU* LEACH URANIUM EXTRACTION LICENSE APPLICATIONS

AGENCY: U.S. Nuclear Regulatory Commission

ACTION: Notice of availability

SUMMARY: The U.S. Nuclear Regulatory Commission (NRC) has developed a Standard Review Plan (NUREG–1569) which provides guidance for staff reviews of applications to develop and operate uranium in situ leach facilities. Under the provisions of Title 10 of the Code of Federal Regulations, Part 40 (10 CFR Part 40), Domestic Licensing of Source Material, an NRC Materials License is required to conduct uranium recovery by *in situ* leach extraction techniques. Applicants for a new license and operators seeking an amendment or renewal of an existing license are required to provided detailed information on the facilities, equipment, and procedures used in the proposed activities. In addition, the applicant for a new license also provides an Environmental Report that discusses the effects of proposed operations on the health and safety of the public and assesses impacts to the environment. For amendment or renewal of an existing license, the original Environmental Report is supplemented, as necessary. This information is used by the NRC staff to determine whether the proposed activities will be protective of public health and safety and the environment and to fulfill NRC responsibilities under the National Environmental Policy Act (NEPA). The purpose of the Standard Review Plan (NUREG–1569) is to provide the NRC staff with guidance on performing reviews of information provided by the applicant, and to ensure a consistent quality and uniformity of staff reviews. Each section in the review plan provides guidance on what is to be reviewed, the basis for the review, how the staff review is to be accomplished, what the staff will find acceptable in a demonstration of compliance with the regulations, and the conclusions that are sought regarding the applicable sections in 10 CFR Part 40, Appendix A. NUREG–1569 is also intended to improve the understanding of the staff review process by interested members of the public and the uranium recovery industry. The review plan provides general guidance on acceptable methods for compliance with the existing regulatory framework. As described in an NRC white paper on risk-informed, performance-based regulation (SECY–98–144), however, the applicant has the flexibility to propose other methods as long as it demonstrates how it will meet regulatory requirements.

A draft of NUREG–1569 was issued in October 1997, and subsequently revised to reflect responses to public comments, and the results of Commission policy decisions affecting uranium recovery issues described in NRC Regulatory Issue Summary 2000–23, dated November 30, 2000. On February 5, 2002 (FR5347), the NRC made the revised second draft of NUREG–1569 available for a 75-day public comment.

In preparing the final version of NUREG–1569, the NRC staff carefully reviewed and considered more than 750 written comments received by the close of the public comment period on April 22, 2002. To simplify the analysis, the NRC staff grouped all comments into the following major topic areas:

(1)   Editorial and Organizational Comments (322 comments)
(2)   Policy Issues (including administrative, quality assurance, and surety/financial issues) (103 comments)

(3)   Ground water (123 comments)
(4)   Operational (47 comments)
(5)   Health Physics (78 comments)
(6)   Monitoring (55 comments)
(7)   Environmental aspects related to NRC responsibilities under NEPA (40 comments)

The following provides a more detailed discussion of the NRC evaluation of the major topic areas and the NRC responses to comments.

1.   Editorial and Organizational Comments

Issue: The standard review plan has a number of redundancies and editorial errors.

Comment. Several commenters identified editorial concerns, text omissions, or areas where the organization of the standard review plan could be improved. Most of the organizational comments addressed perceived redundancies in the standard review plan or opportunities to streamline the style. Most editorial comments addressed inconsistent terminology, identified typographical and grammatical mistakes, or questioned the accuracy of reference documents.

Response. NUREG–1569 is structured consistent with NRC practice for standard review plan style and format. While this style and format may be considered complex or redundant by some commenters, no substantive changes have been made. This will preserve consistency with other NRC standard review plans. The commenters have provided numerous suggestions for improving the readability and clarity of the review plan. Editorial comments on inconsistent terminology, typographical and grammatical mistakes, or the accuracy of reference documents were accepted and incorporated in preparing the final standard review plan, as appropriate. The individual editorial comments are not addressed in this comment summary document.

An appendix (Effluent Disposal at Licensed *In Situ* Leach Uranium Extraction Facilities) was deleted since the guidance therein was superseded by SECY–99–013 which provided staff with direction on classification of liquid wastes at these facilities.

Issue: There is sometimes a lack of agreement between the topics to be reviewed and the corresponding acceptance criteria.

Comment: Commenters stated that in several review plan sections, the areas of review identified at the beginning of the section did not correspond well to the acceptance criteria that would be used to make the evaluation findings.

Response: The staff concurs with this comment. NUREG–1569 was edited to provide correspondence among areas of review, review methods, acceptance criteria, and evaluation findings in each section.

Issue: Chapter 5 (Operations) of the standard review plan has many editorial and technical discrepancies.

Comment: Several commenters identified editorial and technical concerns with Chapter 5 of the draft standard review plan. In some cases, the editorial problems may have made the regulatory guidance difficult to implement.

Resolution: The staff concurs with the commenters. Chapter 5 was rewritten to incorporate editorial and regulatory guidance improvements. The separate section on record keeping and reporting was combined with the section on the management control program to more closely match Regulatory Guide 3.46.1 (Standard Format and Content of License Applications, Including Environmental Reports, for *In Situ* Uranium Solution Mining). Editorial comments are not addressed individually in this comment summary document except where they have particular impact on the standard review plan.

Issue: Additional clarifying or background information should be included in NUREG–1569.

Comment: Several commenters suggested that specific additional information related to proceedings for a given site or that would provide general background information on *in situ* uranium extraction techniques and hazards be included.

Resolution: The NRC has elected not to include the suggested information in NUREG–1569 because the standard review plan is not written for application to a specific site, and general information is available in other references on *in situ* uranium extraction operations.

2. Policy Issues (Including Administrative, Quality Assurance, and Surety/Financial Issues)

Issue: NUREG–1569 attempts to apply a risk-informed, performance-based regulatory philosophy without a regulatory basis for doing so.

Comment: Commenters, while noting that risk-informed, performance-based regulatory philosophies could be applied to *in situ* leach uranium extraction facilities, argued that no regulatory basis exists for implementing such philosophies. The commenters stated that 10 CFR Part 40 should be modified to incorporate risk-informed, performance-based regulatory concepts before the associated standard review plan is modified in that way, because standard review plans are not to be used to promulgate regulatory policy. Commenters also stated that the NRC should not expect license applicants to conduct the accident analyses; consequence evaluations; and probability determinations associated with risk-informed, performance-based regulation. Finally, the commenters argued that the risk-informed, performance-based approach presented in NUREG–1569 was too cursory, contained undefined terms, assumed the existence of a facility change mechanism, and that the review plan contained highly prescriptive acceptance criteria.

Response: The NRC agrees that standard review plans cannot be used to promulgate regulatory requirements, and has no intent to do so using NUREG–1569. In related action, the Commission considered promulgating a new regulation (10 CFR Part 41) that would specifically address regulatory requirements for *in situ* leach uranium extraction facilities and that would formally incorporate risk-informed, performance-based regulatory philosophies. However, considering feedback from the uranium extraction industry and other stakeholders, and taking into account the economic status of the uranium extraction industry, the Commission determined that rulemaking was not an appropriate action at this time. Instead, in making this decision, the Commission directed the staff to update its regulatory guidance related to *in situ* leach uranium extraction facilities, and in so doing, to provide guidance on use of risk-informed, performance-based regulatory philosophies. NUREG - 1569 incorporates this direction from the Commission. It outlines risk-informed, performance-based approaches that staff reviewers may apply to *in situ* leach uranium extraction facilities that are also consistent with existing NRC regulations at 10 CFR Part 40.

In NUREG/CR–6733 (A Baseline Risk-Informed, Performance-Based Approach for *In Situ* Leach Uranium Extraction Licensees) the staff presents analyses of *in situ* leach uranium extraction facility operations and accidents that consider both likelihood of occurrence and consequence (and therefore, risk). The analyses in NUREG–6733 are conservative and demonstrate that *in situ* leach uranium extraction facilities operated with properly trained workers and effective emergency response procedures generally pose low levels of radiologic risk. The staff considers analyses similar to, or based on, those in NUREG–6733 to be an appropriate basis for licensee safety analyses. NUREG–1569 is not intended to require applicants to prepare complex accident analyses, consequence evaluations, and probability determinations. However, site-specific conditions and circumstances must be addressed in any application.

For several years, the NRC staff has been approving *in situ* leach uranium extraction facility license renewals that incorporate a performance-based license condition that provides a facility change mechanism using a Safety and Environmental Review Panel. This accepted practice is continued in NUREG–1569.

Finally, the staff has not attempted to implement overly prescriptive acceptance criteria in NUREG–1569. Rather, standard practices that have been found acceptable in demonstrating compliance at *in situ* leach uranium extraction facilities have been placed in the standard review plan as one approach that the staff may use in determining compliance. The introduction to 10 CFR Part 40, Appendix A, allows applicants to propose alternate methods to demonstrate compliance, and the staff will review any such alternate methods that are submitted.

NUREG–1569 has been edited to remove inconsistent use of terms or undefined terms. Where useful, acceptance criteria have been modified to be less prescriptive. However, risk-informed, performance-based approaches to determining compliance have been incorporated in the standard review plan to the extent consistent with existing regulations.

Issue: Standard review plan guidance with respect to overlapping jurisdiction is not adequate.

Comment: Commenters were concerned that NUREG–1569 did not provide sufficiently clear guidance on coordinating license application reviews with federal and state agencies. Commenters also stated that NRC should accept state guidelines in conducting reviews.

Response: NUREG–1569 implements Commission direction in SECY–99–013 regarding ground-water issues at *in situ* leach uranium extraction facilities. While this direction requires the staff to determine the extent to which it can rely on the U.S. Environmental Protection Agency's (EPA) Underground Injection Control program and to work to implement agreements with appropriate states on these issues, it does not suggest that the NRC broadly accept state guidelines. As appropriate, minimizing dual regulation and implementing agreements with

4

affected states remains an objective of the NRC, and interactions with the EPA and the states continue on these issues. The review plan has been revised to clarify this intent.

Issue: The standard review plan directs the staff to inappropriately seek disclosure of an applicant's primary corporate internal costs.

Comment: Commenters argued that corporate internal costs such as capital costs of land acquisition and improvement, capital costs of facility construction, and other operating and maintenance costs addressed in the draft standard review plan were not appropriate for staff review. The commenters suggested that only the forecast costs for plant decommissioning and site reclamation should be examined by the staff.

Resolution: The staff agrees with the commenters. The standard review plan has been revised to remove guidance that the staff examine costs outside of those associated with plant decommissioning and site reclamation.

Issue: NRC is exceeding its legal authority by requiring that a determination be made that a proposed licensing action is appropriate prior to allowing construction to proceed.

Comment: The Executive Summary to NUREG–1569 states that "beginning construction of process facilities, well fields, or other substantial actions that would adversely affect the environment of the site, before the staff has concluded that the appropriate action is to issue the proposed license, is grounds for denial of the application." The commenter disagrees with the "sweeping nature" of this statement and asserts that NRC has no jurisdiction over wells in an exempted aquifer until lixiviant injection begins.

Response: The NRC considers this statement to be consistent with the requirements of 10 CFR 40.32(e) and believes it to be appropriate for the agency's responsibilities to protect public health and safety and the environment. The license applicant should not conduct actions with a potential for adverse impacts prior to the NRC completing its safety evaluation and environmental assessment.

3.   Ground Water

Issue: Some acceptance criteria for ground-water protection seem overly prescriptive or inconsistent with current practices at specific *In situ* leach uranium extraction facilities.

Comment: Several comments pertained to the use of examples of acceptable methods and approaches cited in the various acceptance criteria for ground-water protection. These comments expressed concern that the examples cited were not consistent with current practices at some *in situ* leach uranium extraction facilities. For example, several comments stated that the examples of acceptable methods for conducting mechanical integrity tests on injection wells are not consistent with methods currently employed or with state-approved practices.

Response: Examples of acceptable practices cited in the review plan acceptance criteria for ground-water protection were obtained from operations plans of currently operating *in situ* leach uranium extraction facilities. These examples refer to methods used to implement ground-water protection requirements that have been considered acceptable in past NRC licensing reviews. The NRC recognizes that an optimal approach to ground-water protection at one facility is not necessarily applicable or appropriate at all *in situ* leach uranium extraction facilities. As stated in the introduction to NUREG–1569, applicants may take approaches to demonstrating compliance that are different from the acceptance criteria in the standard review plan so long as the staff can make the requisite decisions concerning environmental acceptability and compliance with applicable regulations. Where appropriate, these comments were addressed by modifying text to clarify that the given examples are not prescriptive requirements.

Comment: Several comments recommended deletion of constituents from the list of typical baseline water quality indicators in Table 2.7.3-1 of NUREG–1569. In a specific example, a rationale was provided for eliminating radium-228 from the list of baseline water quality indicators to be sampled in each new well field.

Response: The rationales provided by the commenters for elimination of certain chemical constituents from the list of typical baseline water quality indicators are not necessarily applicable for all *in situ* leach uranium extraction facilities. A licensee may provide the rationale for the exclusion of water quality indicators in a license application or amendment request if operational experience or site-specific data demonstrate that concentrations of constituents such as radium-228 are not significantly affected by *in situ* leach operations. NRC reviewers will determine whether proposed exclusions are justified by the information provided. No changes to Table 2.7.3-1 were made for the final standard review plan.

Comment: Two commenters pointed out an apparently new policy that an excursion of lixiviant solutions will be deemed to have occurred if any single excursion indicator exceeds its upper control limit by 20 percent, where previous guidance considered an excursion to have occurred only when two or more excursion indicators exceed their upper control limits by any amount.

Response: Acceptance criterion (5) in Section 5.7.8.3 of NUREG–1569 was revised by deleting the statement regarding a single excursion indicator exceeding its upper control limit by 20 percent for determination of when an excursion has occurred. However, the same acceptance criterion retains the requirement that corrective action for an excursion is deemed complete when all excursion indicators are below their respective upper control limits, or when no single indicator exceeds its control limit by more than 20 percent. Ideally, corrective action for an excursion would be to restore all indicators to below their upper control limits. However, in the past, corrective action has been considered acceptable when a monitor well no longer meets the criteria for being on excursion status. Excursion status criteria allow one indicator to be above the respective upper control limit. However, once an excursion has occurred, the reduction in concentrations of indicator constituents by corrective action may not occur at the same rate. Therefore, corrective action may be terminated prematurely if one of two indicators are brought below upper control limits while another remains substantially above its control limit.

Issue: The NRC is unduly concerned with protection of ground water in aquifers where exemptions have been obtained from the requirements of the Safe Drinking Water Act.

Comment: Several commenters took exception to Acceptance Criterion (4) in Section 6.1.3 of the draft standard review plan, which states that the primary goal for restoration of well fields, following uranium extraction, is to return each well field to its pre-operational baseline water quality conditions. The commenters correctly pointed out that EPA requirements for the Underground Injection Control program result in the uranium production zones being classified as "Exempted Aquifers." This means they are not considered a potential source of drinking water and, therefore, are not subject to requirements of the Safe Drinking Water Act.

Response: Acceptance Criterion (4) of Section 6.1.3 in the standard review plan was revised to clarify that the goal of ground-water restoration at *in situ* leach uranium extraction facilities is to protect present or potential future sources of drinking water outside of the exempted production zone. Generally, if water quality within the production zone is restored to the pre-operational baseline water quality, then protection of water resources outside the exempted zone is assured. Hence, restoration to pre-operational conditions is considered a primary goal whenever degradation of water outside of the exempted zone is a possibility. It is recognized, however, that restoration to pre-operational baseline conditions may not be practicable or feasible, owing to geochemical changes in the production zone during operations. Hence, applicants may propose secondary standards for monitored constituents that are protective of water resources outside of the exempted zone. This has also been clarified in the final standard review plan.

4. Operations

Issue: It is unclear which hazardous chemicals have the potential to impact safety at *in situ* leach uranium extraction facilities.

Comment: Some commenters expressed concern that the standard review plan addressed hazardous chemicals that were not realistic concerns at *in situ* leach uranium extraction facilities.

Response: In 10 CFR Part 40, Appendix A, regulations implement EPA Standards at 40 CFR Part 192, as required by law. Specifically, 10 CFR Part 40, Appendix A, Criterion 13 identifies those hazardous constituents for which standards must be set and complied with if the specific constituent is reasonably expected to be in, or derived from, the byproduct material, and has been detected in ground water. At the same time, the introduction to 10 CFR Part 40, Appendix A allows applicants to submit alternative proposals for meeting the requirements that take into account local or regional conditions. 10 CFR Part 40, Appendix A, Criterion 13 also notes that the Commission does not consider the subsequent list of hazardous constituents to be exhaustive. In summary, NUREG–1569 reflects the regulatory requirements but also allows the reviewer to consider any demonstration presented by an applicant that addresses the potential hazardous constituents at a specific site.

Issue: The responsibilities of the Safety and Environmental Review Panel are not well defined.

Comment: Various commenters stated that the responsibilities of the Safety and Environmental Review Panel, and their authority to authorize changes without a license amendment were either not clear or had no regulatory basis.

Resolution: The staff agrees that clarification of Safety and Environmental Review Panel responsibilities and authorities would facilitate use of the standard review plan. These portions of the plan were rewritten for clarity. However, consistent with a risk-informed, performance-based licensing approach, use of Safety and Environmental Review Panels has been accepted by NRC staff, and an evaluation of their use was left in NUREG–1569.

Issue: NRC is placing inappropriate restrictions on use of potentially hazardous process chemicals at *in situ* leach uranium extraction facilities.

Comment: The commenter refers to NUREG/CR–6733 (A Baseline Risk-Informed, Performance-Based Approach for *In Situ* Leach Uranium Extraction Licensees) and states that the analyses in this document were conservative. The commenter concludes that chemical safety must be based on a realistic analysis of the hazards.

Resolution: The NRC staff interpreted the conclusions from the analyses presented in NUREG/CR–6733 differently from the commenter. NUREG–6733 conducted deliberately conservative analyses for the purpose of evaluating whether risks at *in situ* leach uranium extraction facilities were significant. The conclusion presented in NUREG/CR–6733 for chemical hazards was that licensees should follow design and operating practices published in accepted codes and standards that govern hazardous chemical systems. This recommendation leaves licensees flexibility to establish chemical safety measures appropriate for a specific facility and consistent with good engineering and safety practice. NUREG–1569 places no specific strictures on chemical safety practices at *in situ* leach uranium extraction facilities.

5.   Health Physics

Issue: NRC is requesting information on radiation safety programs that is unnecessary, based on the operational record at *in situ* leach uranium extraction facilities, or is outside NRC licensing authority.

Comment: Some commenters expressed a concern that the NRC was requesting information that is not necessary to fulfill the agency mission of protecting the public health and safety and the environment from the effects of radiation. An example cited was information on radiation safety programs, such as the qualifications of those people proposed for the health physics staff.

Response: The NRC agreed with many of these commenters and revised Chapter 5 of NUREG–1569 to ensure that it is consistent with NRC regulations and regulatory guidance applicable to *in situ* leach uranium extraction facilities.

Issue: NUREG–1569 references regulatory guides that are outdated.

Comment: A number of commenters noted that the standard review plan referenced regulatory guides that have been revised or that are in the process of revision.

Response: The commenters correctly noted that some of the references in the draft standard review plan had been superseded or were in the process of revision. The standard review plan has been edited to reference current guidance. However, NRC has a number of regulatory guides that are being updated, and revised versions may only be referenced when they have been formally approved. This has necessitated retaining reference to some draft regulatory guides.

Issue: NUREG–1569 introduces a new and undefined concept in discussing "control systems relevant to safety."

Comment: Several commenters objected to inconsistent use of terms and a lack of definition for terms related to control systems that may affect safety.

Response: NUREG–1569 was edited to incorporate consistent use of terms, and the term "controls" was defined consistent with other NRC regulatory guidance.

6.   Monitoring

Issue: *In situ* leach uranium extraction facility licensees are not subject to long-term surveillance costs.

Comment: A commenter stated that including long-term surveillance costs in financial surety requirements, as addressed in the draft standard review plan is inappropriate.

Response: NRC staff agrees with the commenter. Reference to long-term surveillance costs has been removed from NUREG–1569.

7.   Comments related to NRC Responsibilities under the National Environmental Policy Act

Issue: NRC is requesting non-radiological information that is outside its area of regulatory authority.

Comment: Many of commenters expressed concern that the NRC was requesting information that is not necessary to fulfill the agency mission of protecting the public health and safety and the environment from the effects of radiation. The areas of concern included information on water quality, air quality, and historical and cultural information.

Response: As a federal agency, the NRC is subject to the NEPA. NEPA requires the NRC to consider impacts to the human environment as a part of its decision making process for licensing actions. The regulations governing NRC implementation of NEPA requirements are at 10 CFR Part 51, Environmental Protection Regulations for Domestic Licensing and Related Regulatory Functions. Guidance to the NRC staff on conducting environmental reviews is also provided in NUREG–1748 "Environmental Review Guidance for Licensing Actions Associated with NMSS Programs." In fulfilling its requirements under NEPA, the NRC routinely prepares an environmental impact assessment when evaluating applications for new materials licenses or amendments to such licenses. Areas of potential environmental impact that are investigated include water availability and quality, air quality, historical and cultural resources, ecology, aesthetic resources, socioeconomic effects, and environmental justice. In preparing its environmental impact assessment under NEPA, it is necessary for NRC to establish background conditions for the affected area. This may require collection of data over a larger geographic area than the licensed area, as well as collection of data in technical and sociological areas that are beyond the traditional scope of radiation safety assessments. The commenters noted that detailed environmental impact assessments may not be necessary for all licensing actions, such as license amendment requests that may be minor in scope or short in duration. The text of the review plan has been modified to clarify those situations where NRC has traditionally performed a detailed environmental impact assessment, but the NRC necessarily reserves the right to determine the nature of the assessment on a site-specific basis in accordance with the requirements of 10 CFR Part 51.

Issue: The standard review plan inappropriately examines corporate financial information in evaluating the socioeconomic effects in cost-benefit analyses.

Comment: A number of commenters noted that the standard review plan examines detailed internal corporate financial data as part the review of cost-benefit analyses for a licensing action. The commenters expressed concern that this information was proprietary and beyond the scope of information necessary for an evaluation of the socioeconomic impact of a facility.

Response: The commenters correctly noted that some of the information identified in the draft standard review plan was beyond the scope of information typically required for cost-benefit analyses. The text of the standard review plan has been revised to eliminate requests for proprietary corporate financial information and to clarify the purpose and use of the financial information that is addressed in the standard review plan.

Issue: Commenters questioned whether the standard review plan applies to facilities planned for private land as well as those on public land.

Comment: Several commenters expressed uncertainty as to whether the review methods and acceptance criteria developed in the standard review plan were also applicable to *in situ* leach facilities wholly located on private lands.

Response: The NRC must consider the environmental impacts of activities on both private and public lands to meet its responsibilities under NEPA, particularly with regard to assessment of direct, indirect, and cumulative impacts of proposed actions. The specific information to be provided by a licensee, and the level of the NRC staff review, will be determined on a site-specific basis considering the nature of the proposed action. The standard review plan is general guidance to the staff on the type of information that is commonly acceptable for evaluating the environmental impact of a proposed licensing action. Consistent with the NRC risk-informed, performance-based licensing philosophy, licensees may use compliance demonstration methods different from those presented in the standard review plan so long as the staff can determine whether public health and safety and the environment are protected. The standard review plan text has been revised for clarity, but it has not been changed to reflect different approaches for facilities operating on private and public lands.

Issue: Licensees should not be required to choose the alternative that has the least impact on the environment.

Comment: Several commenters expressed concern that the standard review plan requires a licensee or applicant to select the alternative that has the least impact on the environment, or requires that NRC use license conditions to mitigate adverse environmental impacts that are deemed outside the scope of NRC responsibilities.

<u>Response</u>: The NRC agrees that while NEPA requires the agency to identify a preferred alternative, it does not require that the alternative with the least impact on the environment be selected. However, if an environmental impact statement (EIS) is necessary for a proposed action, NEPA requires that all reasonable alternatives be evaluated and that the environmentally preferable alternative be identified in the final EIS. NUREG–1569 does not require the applicant or licensee to select the most environmentally benign alternative. As guidance to the NRC staff, the standard review plan asks the reviewers to determine whether the choice of a particular uranium recovery method has been adequately justified and whether different techniques and processes were evaluated as part of this justification. The standard review plan also directs the staff is to evaluate the bases and rationales used by an applicant in evaluating and ranking alternatives.

As stated in Council on Environmental Quality regulations (40 CFR 1502.16), in preparing an EIS, federal agencies are to identify all reasonable mitigation measures that can offset the environmental impacts of a proposed action, even if they are outside the jurisdiction of the lead agency. These mitigation measures are intended to avoid, minimize, rectify, reduce, or compensate for significant impacts of a proposed action. If an environmental assessment identifies potentially significant impacts that can be reduced to less-than-significant levels by mitigation, an agency may issue a mitigated finding of no significant impact (FONSI). In the case of a mitigated FONSI, the mitigation measures should be specific and tangible, such as may be stated as license conditions. The standard review plan states that NRC has responsibilities under NEPA to identify and implement measures to mitigate adverse environmental impacts of the proposed action.

ABSTRACT

A U.S. Nuclear Regulatory Commission source and byproduct materials license is required to recover uranium by *in situ* leach extraction techniques, under the provisions of Title 10 U.S. Code of Federal Regulations, Part 40 (10 CFR Part 40), "Domestic Licensing of Source Material." An applicant for a research and development or commercial-scale license, or for the renewal or amendment of an existing license, is required to provide detailed information on the facilities, equipment, and procedures used and an environmental report that discusses the effects of proposed operations on the health and safety of the public and on the environment.

The standard review plan is prepared for the guidance of staff reviewers, in the Office of Nuclear Material Safety and Safeguards, in performing safety and environmental reviews of applications to develop and operate uranium *in situ* leach facilities. It provides guidance for new license applications, renewals, and amendments. The principal purpose of the standard review plan is to assure the quality and uniformity of staff reviews and to present a well-defined base from which to evaluate changes in the scope and requirements of a review.

The standard review plan is written to cover a variety of site conditions and facility designs. Each section is written to provide a description of the areas of review, review procedures, acceptance criteria, and evaluation findings. However, for a given application, the staff reviewers may select and emphasize particular aspects of each standard review plan section, as appropriate for the application.

Paperwork Reduction Act Statement

The information collections contained in this NUREG are covered by the requirements of 10 CFR Part 40, which were approved by the Office of Management and Budget, approval number 3150-0020.

Public Protection Notification

CONTENTS

CONTENTS (continued)

CONTENTS (continued)

CONTENTS (continued)

CONTENTS (continued)

CONTENTS (continued)

CONTENTS (continued)

CONTENTS (continued)

CONTENTS (continued)

FIGURES

TABLES

INTRODUCTION

A U.S. Nuclear Regulatory Commission (NRC) source and byproduct material license is required under the provisions of Title 10 of the U.S. Code of Federal Regulations, Part 40 (10 CFR Part 40), Domestic Licensing of Source Material, to recover uranium by *in situ* leach techniques. The licensing process for Part 40 licenses is pictured in Figure 1. NRC authority to regulate *in situ* leach facilities comes from the Atomic Energy Act of 1954, as amended, and the Uranium Mill Tailings Radiation Control Act of 1978, as amended. Specific requirements for *in situ* leach facilities are taken from 10 CFR Part 40, Appendix A criteria. The specific sections in this standard review plan that address these criteria are shown in Appendix B of the review plan. Although the National Environmental Policy Act of 1969 does not provide NRC with any additional authority, it does reinforce NRC authority found in the organic statutes by obligating NRC to evaluate both radiological and nonradiological environmental impacts for NRC-licensed sites. Also the National Environmental Policy Act , as interpreted by the courts, requires NRC to mitigate environmental impacts resulting from Agency actions, to the extent possible, through its licensing. Therefore, NRC can also condition commitments made by applicants to mitigate such environmental impacts.

An applicant for a new operating license, or for the renewal or amendment of an existing license, is required to provide detailed information on the facilities, equipment, and procedures to be used and to submit an environmental report that discusses the effect of proposed operations on public health and safety and the impact on the environment as required by 10 CFR 51.45, 51.60, and 51.66. This information is used by NRC staff to determine whether the proposed activities will be protective of public health and safety and will be environmentally acceptable. General provisions for issuance, amendment, transfer, and renewal of licenses are described in 10 CFR Part 2, Subpart A. General guidance for filing an application and for producing an environmental report is provided in 10 CFR 40.31, Application for Specific Licenses, and in 10 CFR Part 51, Environmental Protection Regulations for Domestic Licensing and Related Regulatory Functions, respectively.

The purpose of this standard review plan is to provide the NRC staff in the Office of Nuclear Material Safety and Safeguards with specific guidance on the review of applications for *in situ* leach facilities. The standard review plan complements Regulatory Guide 3.46, Standard Format and Content of License Applications, Including Environmental Reports for *In Situ* Uranium Solution Mining (NRC, 1982) which is guidance to applicants and licensees on an acceptable format and contents for a license application. Sections of this standard review plan are keyed to sections in Regulatory Guide 3.46 (NRC, 1982). Applicants should use Regulatory Guide 3.46 (NRC, 1982) as guidance in preparing their applications. Information in this standard review plan will be used by the Office of Nuclear Material Safety and Safeguards staff in the review of applications for new facilities, renewals, and amendments.

Throughout the remainder of this standard review plan, "application" is synonymous with license application, renewal, or amendment. The principal purpose of the standard review plan is to ensure a consistent quality and uniformity in NRC staff reviews. Each section in this standard review plan provides guidance on what is to be reviewed, the basis for the review, how the staff review is to be accomplished, what the staff will find acceptable in a demonstration of compliance with the regulations, and the conclusions that are sought regarding the applicable sections in Title 10 of the U.S. Code of Federal Regulations. In general, *in situ* leach

Introduction

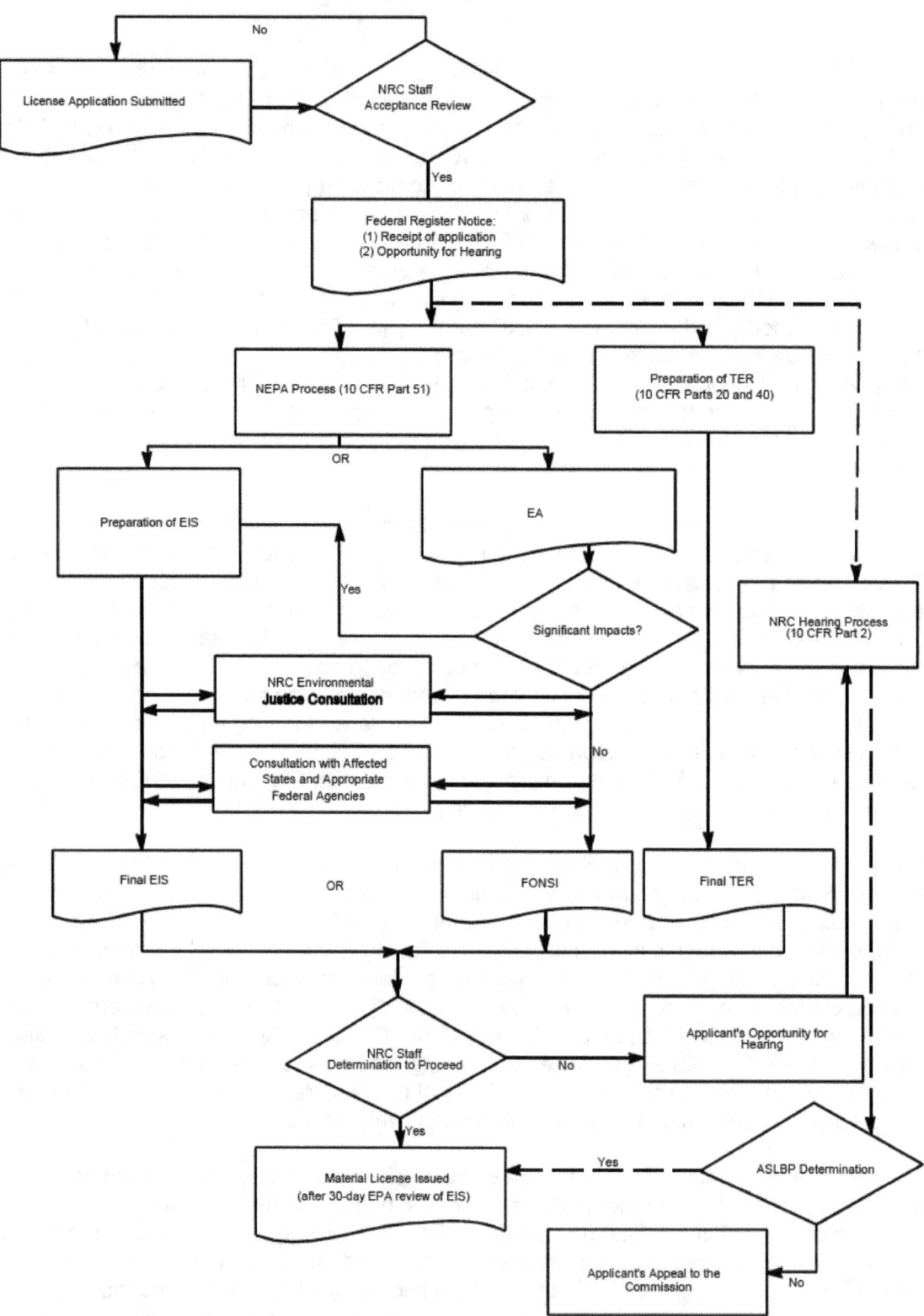

Figure 1. Licensing Process for 10 CFR Part 40 Licenses

operations are much more environmentally benign than conventional mining and milling and pose lower risk of occupational harm. Still, the NRC staff must determine if operations will be conducted in an environmentally acceptable manner and in compliance with applicable regulations. The detailed review procedures and acceptance criteria are intended to assist the Office of Nuclear Material Safety and Safeguards staff in making the necessary findings in an effective and efficient manner. General information regarding procedures for environmental reviews for licensing actions and guidance for the preparation of environmental assessments is available in NUREG–1748, "Environmental Review Guidance for Licensing Actions Associated with NMSS Programs" (NRC, 2001).

This standard review plan is intended to cover only those aspects of the NRC regulatory mission related to the licensing of an *in situ* leach facility. As such, the standard review plan helps focus the staff review on determining if a facility can be constructed and operated in compliance with the applicable NRC regulations. The standard review plan is also intended to make information about regulatory matters widely available and to improve communications and understanding of the staff review process by interested members of the public and the uranium recovery industry.

For amendments, the focus of the review should be on the changes proposed in the amendment (see Appendix A for guidance for reviewing historical aspects of site performance). Reviewers should not review other previously accepted actions if they are not part of the amendment unless the review of the amendment package identifies problems with other aspects of facility operation.

For renewals, the licensee need only submit information containing changes from the currently accepted license. As for amendments, the staff reviews should focus on those aspects of facility operation that are different from what is in the current license. The licensee need not resubmit a complete application covering all aspects of facility operation. Reviewers should analyze the inspection history and operation of the site to see if any major problems have been identified over the course of the license term and should review changes to operations from those currently found acceptable (see Appendix A). If the changes are found to be acceptable, then the license is acceptable for renewal.

For license amendments and renewals, the operating history of the facility is often a valuable source of information concerning the adequacy of site characterization, the acceptability of radiation protection and monitoring programs, the success of and adherence to operating procedures and training programs, and other data that may influence the staff's determination of compliance. Appendix A to the standard review plan provides guidance for review of these historical aspects of facility performance.

The products that will be prepared by the staff to document the review will be a technical evaluation report, and an environmental assessment with a finding of no significant impact to meet requirements under the National Environmental Policy Act. Preparation of an environmental assessment is required under the provisions of 10 CFR 51.20 unless (i) the staff finds, based on the environmental assessment, that NRC needs to prepare an environmental impact statement; (ii) an environmental impact statement is needed by another federal agency also involved in the action as a cooperating agency; (iii) an environmental impact statement would be needed because of controversy at the site, or (iv) the action is categorically excluded

Introduction

from the necessity to prepare an environmental assessment by 10 CFR 51.22. Different sections of this standard review plan refer either to a technical evaluation report, an environmental assessment, or both. Table 1 identifies which sections apply to a technical evaluation report and which to an environmental assessment. Details on the NRC National Environmental Policy Act process are contained in NUREG–1748, "Environmental Review Guidance for Licensing Actions Associated with NMSS Programs" (NRC, 2001).

It is important to note that the acceptance criteria laid out in this standard review plan are for the guidance of NRC staff responsible for the review of applications to operate *in situ* leach facilities. Review plans are not substitutes for the Commission's regulations, and compliance with a particular standard review plan is not required. This standard review plan provides descriptions of methodologies that have been found acceptable for demonstrating regulatory compliance. Methods and solutions different from those set out in the standard review plan will be acceptable if they provide a basis for the findings requisite to the issuance or continuance of a license by NRC.

General Review Procedure

A licensing review is not intended to be a detailed evaluation of all aspects of facility operations. Specific information about implementation of the program outlined in an application is obtained through NRC review of procedures and operations done as part of the inspection function. A definition of the differences between licensing reviews and inspections is provided in Figure 2.

The general licensing process is outlined in the flow diagram provided in Figure 1. An *in situ* leach source and byproduct material application may be denied or rejected under specific instances during the review process. Beginning construction of process facilities, well fields, or other substantial actions that would adversely affect the environment of the site, before the staff has concluded that the appropriate action is to issue the proposed license, is grounds for denial of the application [10 CFR 40.32(e)]. The applicant's failure to demonstrate compliance with requirements [10 CFR 40.31(h)], or refusal or failure to supply information requested by the staff to complete the review (10 CFR 2.108) is also grounds for denial of the application.

Changes to existing licensed activities and conditions require the issuance of an appropriate license amendment. An application for such an amendment should describe the proposed changes in detail and should discuss the likely consequences of any environmental and health and safety impacts. Amendment requests should be reviewed using the appropriate sections of this document for guidance. Appendix A to this standard review plan provides guidance for examining the historical aspects of facility operations that may be useful for conducting such amendment reviews.

In conducting these evaluations, the reviewer shall consider the technical evaluations conducted by a state or another federal agency with authorities overlapping those of the NRC. Ground-water compliance and protection reviews are the primary technical areas impacted by overlapping authorities. The desired outcome is to identify any areas where duplicative NRC reviews may be reduced or eliminated. The NRC staff must make the necessary evaluations of compliance with applicable regulations for licensing the facility. However, the reviewer may, as

		Applicable to Technical Evaluation Report	Applicable to Environmental Assessment
Section	**Title**		
1.0	PROPOSED ACTIVITIES	X	X
2.0	SITE CHARACTERIZATION	X	X
2.1	Site Location and Layout	X	X
2.2	Uses of Adjacent Lands and Waters	X	X
2.3	Population Distribution	X	X
2.4	Historic, Scenic, and Cultural resources		X
2.5	Meteorology	X	X
2.6	Geology and Seismology	X	X
2.7	Hydrology	X	X
2.8	Ecology	X	X
2.9	Background Radiological Characteristics	X	X
2.10	Background Non-Radiological Characteristics	X	X
3.0	DESCRIPTION OF PROPOSED FACILITY	X	X
3.1	In Situ Leaching Process and Eqiupment	X	X
3.2	Recovery Plant Equipment	X	X
3.3	Instrumentation and Control	X	X
4.0	EFFLUENT CONTROL SYSTEMS	X	X
4.1	Gaseous and Airborne Particulates	X	X
4.2	Liquids and Solids	X	X
4.3	Contaminated Equipment	X	X
5.0	OPERATIONS	X	
5.1	Corporate Organization and Administrative Procedures	X	

Table 1. Identification of Sections Applicable to a Technical Evaluation Report or an Environmental Assessment

Table 1. Identification of Sections Applicable to a Technical Evaluation Report or an Environmental Assessment (continued)

Introduction

Section	Title	Applicable to Technical Evaluation Report	Applicable to Environmental Assessment
5.2	Management Control Program	X	
5.3	Management Audit, Inspection, and Record-keeping Program	X	
5.3.1	Management Audit, and Internal Inspection Program	X	
5.3.2	Recordkeeping and Record Retention	X	
5.4	Qualifications for Personnel	X	
5.5	Radiation Safety Training	X	
5.6	Security	X	X
5.7	Radiation Safety Controls and Monitoring	X	
5.7.1	Effluent Control Techniques	X	
5.7.2	External Radiation Exposure Monitoring Program	X	
5.7.3	Airborne Radiation Monitoring Program	X	
5.7.4	Exposure Calculations	X	
5.7.5	Bioassay Program	X	
5.7.6	Contamination Control Program	X	X
5.7.7	Airborne Effluent and Environmental Monitoring Program	X	X
5.7.8	Ground-Water and Surface-Water Monitoring Programs	X	X
5.7.9	Quality Assurance	X	X
6.0	GROUND-WATER QUALITY RESTORATION, SURFACE RECLAMATION, AND PLANT DECOMMISSIONING	X	X
6.1	Plans and Schedules for Ground-Water Quality Restoration		X
6.2	Plans and Schedules for Reclaiming Disturbed Lands		X
6.3	Procedures for Removing and Disposing of Structures and Equipment	X	X
Table 1. Identification of Sections Applicable to a Technical Evaluation Report or an Environmental Assessment (continued)			

Section	Title	Applicable to Technical Evaluation Report	Applicable to Environmental Assessment
6.4	Procedures for Conducting Post-Reclamation and Decommissioning Radiological Surveys	X	X
6.5	Financial Assessment for Ground-Water Restoration, Decommissioning, Reclamation, Waste Disposal, and Monitoring	X	X
7.0	ENVIRONMENTAL EFFECTS		X
7.1	Site Preparation and Construction		X
7.2	Effects of Operations	X	X
7.3	Radiological Effects	X	X
7.3.1	Exposure Pathways	X	X
7.3.1.1	Exposures from Water Pathways	X	X
7.3.1.2	Exposures from Air Pathways	X	X
7.3.1.3	Exposures from External Radiation	X	X
7.3.1.4	Total Human Exposures	X	X
7.3.1.5	Exposures to Flora and Fauna	X	X
7.4	Non-Radiological Effects		X
7.5	Effects of Accidents	X	X
7.6	Economic and Social Effects of Construction and Operation		X
7.6.1	Benefits		X
7.6.2	Socioeconomic Costs		X
8.0	ALTERNATIVES TO PROPOSED ACTION		X
9.0	COST-BENEFT ANALYSIS		X
10.0	ENVIRONMENTAL APPROVALS AND CONSULTATIONS		X

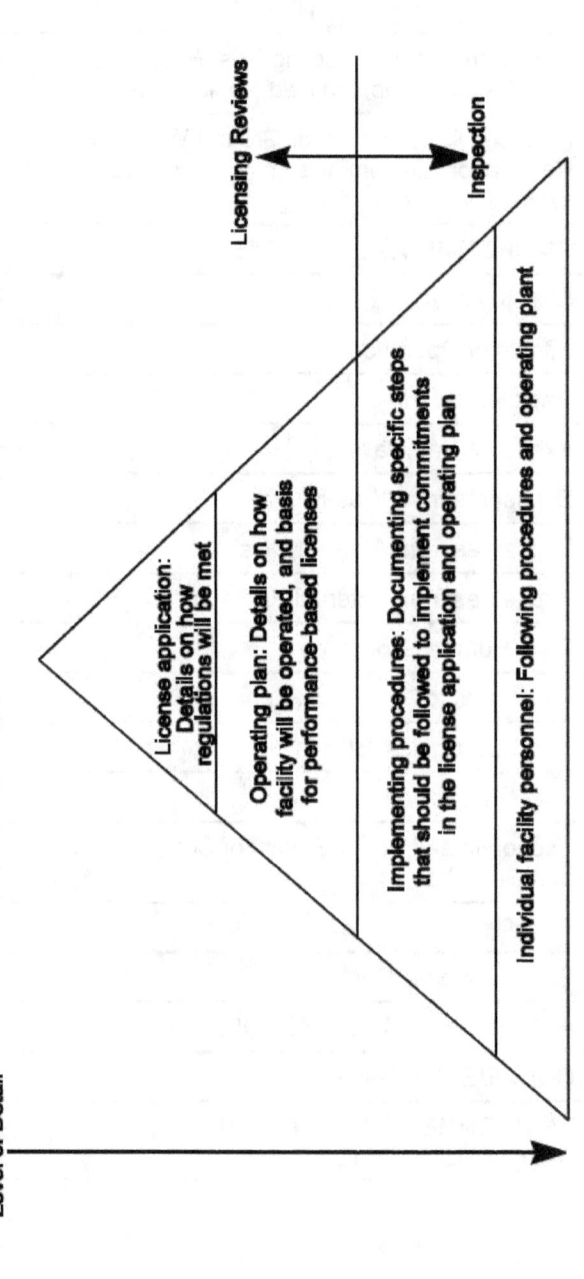

Figure 2. Schematic of NRC Licensing and Inspection Process and Applicability to Different License Documents

appropriate, rely on the applicant's responses to inquiries made by a state or another federal agency to support the NRC evaluation of compliance. The reviewer should make every effort to coordinate the NRC technical review with the state or other federal agency with overlapping authority to avoid unnecessary duplication of effort.

The steps of the application review are described in the following paragraphs.

Acceptance (Administrative) Review Objectives

The staff should conduct an acceptance review of the application, which is an administrative review, to determine the completeness of the information submitted. This review requires a comparison of the submitted information to the information identified in the Standard Format and Content of License Applications, Including Environmental Reports (NRC, 1982). The application will be considered complete for docketing if the information provided is complete, reflects an adequate reconnaissance and physical examination of the regional and site conditions, and provides appropriate analyses and design information to demonstrate that the applicable acceptance criteria will be met. Details for review of the environmental report are also contained in NUREG–1748 (NRC, 2001, Section 6). The staff should complete the acceptance review and transmit the results to the applicant within 30 days of the receipt of the application, along with a projected schedule for the remainder of the review as described in Section 1.1 of the standard review plan. In this transmittal, the staff should identify any additional information needed to make the application complete. Detailed technical questions, although not required, can be included if they are identified during the acceptance review. If the content of the application is acceptable for docketing, the staff should be able to make a finding that the applicable requirements in 10 CFR 40.31 have been met.

Detailed Review Objectives

Following completion of the acceptance review, the staff should conduct a detailed technical review of the application. The results of this review and the basis for acceptance or denial of the requested licensing action are documented by NRC in a technical evaluation report and either an environmental assessment (10 CFR 51.30) if there is a finding of no significant impact, or an environmental impact statement (10 CFR 50.31) if the review indicates that the licensed activity would have a significant impact on the health and safety of the public or on the environment. The detailed review should evaluate the environmental, economic, and technical evidence provided by the applicant to support the ability of the proposed facility to meet applicable regulatory requirements. Details on the NRC National Environmental Policy Act process are contained in NUREG–1748 (NRC, 2001).

Standard Review Plan Organization

The standard review plan is written to address a variety of site conditions and facility designs. Each section provides the complete review procedure and acceptance criteria for all the areas of review pertinent to that section. For any given application, the staff reviewer may select and emphasize particular aspects of each standard review plan section as appropriate for the application. Because of this, the staff may not carry out in detail all of the review steps listed in each standard review plan section in the review of every application.

Introduction

Areas of Review Subsection

This subsection describes the scope of the review (i.e., what is being reviewed). It contains a brief description of the specific technical information and analyses in the application that should be reviewed by each technical reviewer.

Review Procedures Subsection

This subsection discusses the appropriate review technique. It is generally a step-by-step procedure that the reviewer uses to determine whether the acceptance criteria have been met.

Acceptance Criteria Subsection

This subsection delineates criteria that can be applied by the reviewer to determine the acceptability of the applicant compliance demonstration. Because the criteria are based on detailed technical approaches for determining compliance with applicable regulations, they do not routinely reference specific regulations. To include such reference would simply restate the requirements, and would not provide guidance on what is an acceptable method of compliance. The technical bases for these criteria have been derived from 10 CFR Parts 40 and 20, NRC regulatory guides, general design criteria, codes and standards, branch technical positions, standard testing methods (e.g., American Society for Testing and Materials standards), technical papers, and other similar sources. These sources typically include solutions and approaches previously determined to be acceptable by the staff for making compliance determinations for the specific area of review. These acceptance criteria have been defined so that staff reviewers can use consistent and well-documented approaches for review of all applications. Flexibility is provided to enable licensees to achieve the type of operation desired at their facilities. Applicants may take approaches to demonstrating compliance that are different from the acceptance criteria in this standard review plan as long as the staff can make the requisite decisions concerning environmental acceptability and compliance with applicable regulations. However, applicants should recognize that, as is the case for regulatory guides, substantial staff time and effort have gone into the development of these procedures and criteria, and a corresponding amount of time and effort may be required to review and accept new or different solutions and approaches. Thus, applicants proposing solutions and approaches to safety problems or safety-related design issues other than those described in this standard review plan may experience longer review times and NRC requests for more extensive supporting information. The staff is willing to consider proposals for other solutions and approaches on a generic basis, apart from a specific application, to avoid the impact of the additional review time for individual cases.

Evaluation Findings Subsection

This subsection presents general conclusions and findings of the staff that result from review of each area of the application as well as an identification of the applicable regulatory requirements. Conclusions and findings for a specific application and review area are dependent on the site and type of licensing action being considered. For each standard review plan section, a conclusion is included in the technical evaluation report or the environmental assessment/environmental impact statement in which results of the review are published.

These documents contain a description of the review; the basis for the staff findings, including aspects of the review selected or emphasized; where the facility design or the applicant programs deviate from the criteria stated in the standard review plan; and the evaluation findings.

References Subsection

This subsection lists any applicable references.

Standard Review Plan Updates

This standard review plan will be revised and updated periodically as the need arises to clarify the content or correct errors and to incorporate modifications approved by NRC management. Corresponding changes to the Standard Format and Content of License Applications, Including Environmental Reports (NRC, 1982) will be made as required.

References

NRC. NUREG–1748, "Environmental Review Guidance for Licensing Actions Associated with NMSS Programs." Washington, DC: NRC. 2001.

——. Regulatory Guide 3.46, "Standard Format and Content of License Applications, Including Environmental Reports, for *In Situ* Uranium Solution Mining." Washington, DC: NRC, Office of Standards Development. 1982.

1.0 PROPOSED ACTIVITIES

1.1   Areas of Review

The reviewer should examine the summary of the proposed activities for which a license is requested to gain a basic understanding of those proposed activities and the likely consequences of any safety or environmental impact. The staff should review the corporate entities involved; the location of the proposed activities; land ownership; ore-body locations and estimated uranium (U_3O_8) content; proposed solution extraction method and recovery processes; operating plans, design throughput and anticipated annual U_3O_8 production; radiation safety protection estimated schedules for construction, startup, and duration of operations; plans for project waste management and disposal; source and byproduct material transportation plans; plans for ground-water quality restoration, decommissioning, and land reclamation; and surety arrangements covering eventual facility decommissioning, ground-water quality restoration, and site reclamation.

1.2   Review Procedures

The reviewer should determine whether the application provides a sufficiently comprehensive summary of the nature of the facilities, equipment, and procedures to be used in the proposed *in situ* leach activity including the name and location. Reviewers should keep in mind that the development and initial licensing of an *in situ* leach facility is not based on comprehensive information. This is because *in situ* leach facilities obtain enough information to generally locate the ore body and to understand the natural systems involved. More detailed information is developed as each area is brought into production. Therefore, reviewers should verify that sufficient information is presented to reach only the conclusion necessary for initial licensing. However, reviewers should not expect that information needed to fully describe each aspect of a full operation will be available in the initial application. For license renewals and amendment applications, Appendix A to this standard review plan provides guidance for examining facility operations and the approach that should be used in evaluating amendments and renewal applications.

Applications for licenses authorizing commercial-scale operations should rely on results from research and development operations or other operational experience that can be used as a basis to support the proposed processes, operating plans (including plans for ground-water quality restoration), and assessment of the likely consequences of any environmental impact. This does not mean that the applicant needs to develop a research and development facility in order to license a full-scale production plant. Rather it is intended to allow the applicant to rely on available data from research and development facilities, other sites currently operated by the applicant, or sites with similar designs or natural features operated by other licensees. In performing the evaluation, the reviewer should use the data available from these other sources to assess how the proposed site compares with already licensed sites.

Proposed Activities

1.3 <u>Acceptance Criteria</u>

The proposed activities are acceptable if they meet the following criteria:

(1) The application summary of proposed activities includes descriptions of the following items that are sufficient to provide a basic understanding of the proposed activities and the likely consequences of any health, safety, and environmental impact. The content of the introduction is outlined in the "Standard Format and Content of License Applications, Including Environmental Reports, for *In Situ* Uranium Solution Mining" [U.S. Nuclear Regulatory Commission (NRC), 1982].

   (a) Corporate entities involved

   (b) Location of the proposed facilities by county and state, including the facility name

   (c) Land ownership

   (d) Ore-body locations and estimated U_3O_8 content

   (e) Proposed solution extraction method and recovery process

   (f) Operating plans, design throughput, and annual U_3O_8 production

   (g) Estimated schedules for construction, startup, and duration of operations

   (h) Plans for project waste management and disposal

   (i) Plans for ground-water quality restoration, decommissioning, and land reclamation

   (j) Surety arrangements covering eventual facility decommissioning, ground-water quality restoration, and site reclamation

   (k) For license renewals, a summary of proposed changes, a record of amendments since the last license issuance, and documentation of inspection results

(2) Applications for commercial-scale operations include results from research and development operations or previous operating experience as a basis for the proposed processes, operating plans, ground-water quality restoration, and assessment of the likely consequences of any environmental impact.

1.4   Evaluation Findings

If the staff review, as described in this section, results in the acceptance of the summary of the proposed activities, the following conclusions may be presented in the technical evaluation report and in the environmental assessment.
The NRC has completed its review of the summary of the proposed activities at the _____*in situ* leach facility. This review included an evaluation of the methods that will be used to evaluate the proposed activities using the review procedures in standard review plan Section 1.2 and the acceptance criteria outlined in standard review plan Section 1.3.

The applicant has acceptably described the proposed activities at the _____ *in situ* leach facility including (i) corporate entities involved; (ii) location of the proposed facility; (iii) land ownership; (iv) ore-body locations and estimated U_3O_8 content; (v) proposed solution extraction method and recovery process; (vi) operating plans, design throughput, and annual U_3O_8 production; (vii) schedules for construction, startup, and duration of operations; (viii) waste management and disposal plans; and (ix) ground-water quality restoration, decommissioning, and land reclamation plans; (x) surety arrangements covering facility decommissioning, ground-water quality restoration, and site reclamation. For license renewals, the applicant has provided a summary of proposed changes, a record of amendments since the last license issuance, and documentation of inspection results. Applicants for commercial-scale operations have included results from research and development operations or previous operating experience.

Based on the information provided in the application and the detailed review conducted of the summary of the proposed activities at the _____ *in situ* leach facility, the staff concludes that the summary of the proposed activities is acceptable and is in compliance with 10 CFR 40.31, which describes the general requirements for the issuance of a specific license. The summary of proposed activities is acceptable and is in compliance with 10 CFR 51.45, which requires a description of the proposed action sufficient to allow the staff to evaluate the impacts on the affected environment.

1.5   Reference

NRC. Regulatory Guide 3.46, "Standard Format and Content of License Applications, Including Environmental Reports, for *In Situ* Uranium Solution Mining." Washington, DC: NRC, Office of Standards Development. 1982.

2.0 SITE CHARACTERIZATION

2.1 Site Location and Layout

2.1.1 Areas of Review

The staff should review geographic maps, topographic maps, and drawings that identify the site and its location relative to federal, state, county, and other political subdivisions. These should include maps provided to show the location and layout of the proposed facilities, well fields, and all principal structures such as surface impoundments, deep injection wells, recovery plant buildings, exclusion area boundaries and fences, applicant property and leases, and adjacent properties.

The regional location and site layout for the proposed *in situ* leach operations should be reviewed using maps that show the relationship of the site to local water bodies (lakes and streams); geographic features (highlands, forests); geologic features (faults, folds, outcrops); transportation links (roads, rails, airports, waterways); political subdivisions (counties, townships); population centers (cities, towns); historical and archeological features; key species habitat; and non-applicant property (farms, settlements). A contour map of the site showing a plan layout of constructions, significant topographic variations of the site environs, and drainage gradients, should be evaluated.

2.1.2 Review Procedures

The reviewer should establish the validity and completeness of the basic data, to determine that the site location and layout proposed in the application are complete and accurate, and that the site information is sufficient to evaluate the location of the proposed facilities relative to key features and activities. For new applications, the staff should conduct a site visit of the facility, after becoming familiar with the submitted materials, to develop an acceptable familiarization for the review and to verify the general aspects of the submitted materials.

The staff should examine maps and drawings provided in the application and associated environmental reports to determine whether they provide sufficient detail to locate the site regionally relative to local political subdivisions and natural and man-made features and that the maps allow the staff to determine the proposed layout within the existing topography at the site. On a regional scale, the reviewer should examine the location of the facility and all federal, state, County, and local political subdivisions that have a bearing on estimating the environmental impact of the proposed operations. The staff should verify that the total acreage that is owned or leased by the applicant and the portion of that real estate or any adjacent properties that could be affected by site activities have been identified. The reviewer should examine a contour map to determine that the contour intervals and information included on the map are sufficient to show any significant variations in site environs and important drainage gradients. The staff should also determine that the relationship between the site and surface drainage is readily apparent from the provided maps. Likewise, it should be possible to ascertain the likely areas of and effects of site activities on local flora and fauna from the location maps. The staff should determine that the scale and clarity of the maps are adequate to conduct the necessary environmental and safety reviews.

Site Characterization

Reviewers should keep in mind that the development and initial licensing of an *in situ* leach facility is not based on comprehensive information. This is because *in situ* leach facilities obtain enough information to generally locate the ore body and understand the natural systems involved. More detailed information is developed as each area is brought into production. Therefore, reviewers should ensure that sufficient information is presented to reach only the conclusion necessary for initial licensing. However, reviewers should not expect that information needed to fully describe each aspect of all the operations will be available in the initial application.

For license renewals and amendment applications, Appendix A to this standard review plan provides guidance for examining facility operations and the approach that should be used in evaluating amendments and renewal applications.

2.1.3 Acceptance Criteria

The characterization of the site location and layout is acceptable if it meets the following criteria:

(1)   Maps are provided that show geologic features, well fields, and all planned principal structures such as surface impoundments, diversion channels, monitoring wells, deep injection wells, and recovery plant buildings. If detailed information on actual well field design is not available at the time of the initial facility application, the maps show the expected well field locations with an indication that this information is preliminary.

(2)   Any maps previously submitted (e.g., maps from the original application in the case of renewals) are legible, and actual or proposed changes are highlighted.

(3)   Maps are provided that show exclusion area boundaries and fences.

(4)   Maps are provided that show the applicant property and leases and current adjacent properties, including water bodies, forests, and farms, and all federal, state, county, and local political subdivisions.

(5)   Maps are provided that show nearby population centers and transportation links such as railroads, highways, and waterways.

(6)   A topographic map is provided with elevation contours that show the locations of drainage basins and variations in the drainage gradient in the vicinity of the proposed *in situ* leach facility. The specific locations of natural streams and proposed diversion channels, relative to principal structures, should also be provided.

(7)   The proposed *in situ* leach facility is clearly labeled at a scale appropriate to the area being covered (regional and local) and with sufficient clarity and detail to allow identification and evaluation of the proposed *in situ* leach facility. Maps are at an appropriate scale and are clear and readable.

(8)   Data sources are documented in reports such as U.S. Geological Survey open files or existing published maps. If data have been generated by the applicant, the data documentation should include a description of the investigation and data reduction techniques.

(9)   Maps include designation of scale, orientation (e.g., north arrow), and geographic coordinates. In addition to maps, the applicant may provide tabular locations of facilities using universal transverse Mercator coordinates with appropriate Northing and Easting in meters.

2.1.4 Evaluation Findings

If the staff review as described in this section results in the acceptance of the description of the site location and layout, the following conclusions may be presented in the technical evaluation report and in the environmental assessment.

NRC has completed its review of the site characterization information concerned with site location and layout at the _____ *in situ* leach facility. This review included an evaluation using the review procedures in standard review plan Section 2.1.2 and the acceptance criteria outlined in standard review plan Section 2.1.3.

The licensee has acceptably described the site location and layout with appropriately scaled and labeled maps showing site layout, principal facilities and structures, regional location, geology, boundaries, exclusion areas and fences, applicant property including leases and adjacent properties, nearby population centers and transportation links, and topography. References are cited acceptably. Any maps previously submitted (e.g., maps from the original application in the case of renewals) are legible, and actual or proposed changes are highlighted.

Based on the information provided in the application, and the detailed review conducted of the characterization of site location and layout for the _____ *in situ* leach facility, the staff concludes that the information is acceptable and is in compliance with 10 CFR 51.45, which requires a description of the affected environment containing sufficient data to aid the Commission in its conduct of an independent analysis.

2.1.5 References

None.

2.2   Uses of Adjacent Lands and Waters

2.2.1 Areas of Review

The staff should review descriptions of the nature and extent of present and projected land use (e.g., agriculture, sanctuaries, hunting, mining, grazing, industry, recreation, roads), any recent trends or changes in population or industrial patterns, and any other nuclear fuel cycle facilities located or proposed within an 80-km [50-mi] radius of the site.

The staff should also review tables showing, for each of the 22½-degree sectors centered on each of the 16 compass points (i.e., north, north-northeast, etc.), the distances {to a distance of 3.3 km [2 mi]} from the center of the site to the nearest resident and to the nearest site boundary.

The staff review should include the location, nature, and amounts of present and projected surface-and ground-water use (e.g., water supplies, irrigation, reservoirs, recreation, and transportation) within 3.3 km [2 mi] of the site boundary {0.8 km [0.5 mi] for research and development operations} and the present and projected population associated with each use point.

2.2.2 Review Procedures

The reviewer should determine whether the application provides sufficient information on the use of the lands and waters within a 3.3 km [2 mi] distance from the site boundary surrounding the proposed facilities {0.8 km [0.5 mi] for research and development operations} to assess the likely consequences of any impacts of *in situ* leach operations on adjacent properties.

The staff should determine that the application contains the location of residences, ground-water supply wells, surface-water reservoirs, and the estimated use of water in the lands surrounding the site of the proposed facility. Data sources should be referenced. This information should be evaluated to determine whether it is sufficient to delineate the likely impact(s) of the facility, under both normal operating conditions and accidents, on the ground water, surface water, and population (both human and animal) near the site. The reviewer should determine that within 3.3 km [2 mi] from the site boundary, the nature and extent of present and projected water and land use and any other trends or changes in population or industrial patterns have been reported. Any other nuclear fuel cycle facilities located or proposed within an 80-km [50-mi] radius of the site should be identified.

For license renewals and amendment applications, Appendix A to this standard review plan provides guidance for examining historical aspects of facility performance and the approach that should be used in evaluating amendments and renewal applications.

2.2.3 Acceptance Criteria

The characterization of the uses of adjacent lands and waters is acceptable if it meets the following criteria:

(1)   Information is presented in detail sufficient to understand the surrounding land and water uses, such that the likely consequences imposed by *in situ* leach operations can be adequately assessed.

Although the specific requirements may vary from site to site, the general purpose for determining land and water use patterns is to provide supporting data for exposure calculations, cost-benefit analyses, and determinations of air emissions (e.g., dust). A 3.3-km [2-mi] distance from the site boundary is an acceptable area for which land and water use data should be collected. One acceptable method for presenting these data is for the applicant to provide the information requested in the Standard Format and Content of License Applications, Including Environmental Reports (NRC, 1982), Section 2.2. The information presented should include:

(a) Maps showing the locations of nearest residences, ground-water supply wells, and abandoned wells

(b) Types of present and projected (life of facility) water use (e.g., municipal, domestic, agriculture, livestock) and descriptions of the methodology and sources used to develop projections

(c) Present and projected (life of facility) water use estimates, by type, for both ground water and surface water, including present and projected withdrawal, and descriptions of the methodology and sources used to develop projections

(d) For existing ground-water wells, well depth, ground-water elevations, flow rates, drawdown, and a description of the producing aquifer(s)

(e) The locations of abandoned wells and drill holes, including the depth, type of use, condition of closing, plugging procedure used, and date of completion for each well or drill hole within the site area and within 0.4 km [.25 mi] of the well field boundary

(f) Descriptions of the nature and extent of projected land use (e.g., agriculture, recreation, industry, grazing, and infrastructure) and descriptions of the methodology and sources used to develop projections

(g) The location of any other nuclear fuel cycle facilities located or proposed within an 80-km [50-mi] radius of the site

(2) For each of the 22½-degree sectors centered on the 16 cardinal compass points, the information identified in Section 2.2.3 of the Standard Format and Content of License Application, Including Environment Report (NRC, 1982) concerning human residences, nearest site boundary(ies) to residences, surface- and ground-water use, and projected water use, is provided. As described in Section 2.2 of the Standard Format and Content of License Application, Including Environment Report (NRC, 1982), appropriate presentation of the data should include mapped data as appropriate, a tabular summary for each of the 22½-degree sectors centered on the 16 cardinal compass points, and for each, the distance from the center of the site to the site boundary and the nearest residence.

Site Characterization

(3)   Data sources are documented in reports such as U.S. Geological Survey open files or
        existing published reports or maps. If data have been generated by the applicant, the
        data documentation should include a description of the investigations and data
        reduction techniques.

(4)   Maps include designation of scale, orientation (e.g., north arrow), and
        geographic coordinates.

2.2.4   Evaluation Findings

If the staff review as described in this section results in the acceptance of the described uses of
adjacent lands and waters, the following conclusions may be presented in the technical
evaluation report and in the environmental assessment.

NRC has completed its review of the site characterization information concerned with uses of
adjacent lands and waters near the _____ in situ leach facility. This review
included an evaluation using the review procedures in standard review plan Section 2.2.2 and
acceptance criteria outlined in standard review plan Section 2.2.3.

The applicant has acceptably described the present and projected land use, including
residential, commercial, agricultural, industrial, flora and fauna sanctuaries, arboreal, grazing,
recreation (e.g., hunting, swimming, skiing), and infrastructure. Appropriate information on the
location and extent of each use has been provided. In particular, the description and
associated tabulated data of the location, nature, amounts, and population associated with each
use point of present and projected (life of the facility) surface and ground water adjacent to the
site including water supplies, irrigation, reservoirs, recreation, and transportation within at least
3.3 km [2 mi] of the site boundary {0.8 km [0.5 mi] for research and development operations}
are acceptable for determination of likely impacts of the proposed in situ leach facility.
Tabulated data on present and projected water withdrawal rates, return rates, types of water
use (e.g., municipal, domestic, agriculture, and livestock); source, water-use estimates, and
abandoned well locations are acceptable. The applicant has identified and located (or has
noted the absence of) other nuclear fuel cycle facilities located or proposed within an 80-km
[50-mi] radius of the site.

Based on the information provided in the application, and the detailed review conducted of the
characterization of uses of adjacent lands and waters for the _____ in situ leach
facility, the staff concludes that the information is acceptable and is in compliance with
10 CFR 51.45 which requires a description of the affected environment containing sufficient
data to aid the Commission in its conduct of an independent analysis, and 10 CFR Part 40,
Appendix A, Criteria 5B(4) and 5G(3) which provide criteria for identification if underground
sources of drinking water and exempted aquifers and the current uses of ground water.

2.2.5 Reference

NRC. Regulatory Guide 3.46, "Standard Format and Content of License Applications, Including Environmental Reports, for *In Situ* Uranium Solution Mining." Washington, DC: NRC, Office of Standards Development. 1982.

2.3 Population Distribution

2.3.1 Areas of Review

The staff should review population data based on the most recent census, including maps that identify places of significant population grouping, such as cities and towns within an 80-km [50-mi] radius {3.2 km [2 mi] for research and development operations} from the approximate center of projected (life of facility) activities in the format specified in the Standard Format and Content of License Application, Including Environmental Reports (NRC, 1982). For the purposes of environmental justice (see Sections 7.6.1.3) and NUREG–1748 (NRC, 2001) the staff should also examine the distribution of low-income and minority populations based on the most recent census data available. The staff should review the basis for population projections.

In addition, for commercial-scale operations, the staff should review descriptive material giving significant population and visitor statistics of neighboring schools, plants, hospitals, sports facilities, residential areas, parks, *et cetera*, within 3.3 km [2 mi] of the *in situ* leach operations. The review should include appropriate available food production data in kg/yr for vegetables (by type and totals), meat (all types), and milk, and any available future predictions for this production by local governmental, industrial, or institutional organizations within 3.3 km [2 mi] of the site boundary.

2.3.2 Review Procedures

The reviewer should determine that data have been tabulated and presented in pie segments as described in Section 2.3 of the Standard Format and Content of License Application, Including Environmental Reports (NRC, 1982). The basis for population projections should be examined. Recent agricultural production data should be tabulated for vegetables, meat, milk, and other foodstuffs, in addition to predictions for future production by government, industry, or institutions for land within 3.3 km [2 mi] of the site. It is important to ascertain that the most recent census data have been used and that the data presented will support subsequent exposure and dose calculations and risk assessments.

For license renewals and amendment applications, Appendix A to this standard review plan provides guidance for examining facility operations and the approach that should be used in evaluating amendments and renewal applications.

Site Characterization

2.3.3 Acceptance Criteria

The characterization of the population distribution is acceptable if it meets the following criteria:

(1)   Population data including demographic information on minority and low-income populations are provided based on generally accepted sources such as the U.S. Census Bureau, and other federal, state, and local agencies.

(2)   A map of suitable scale is provided that identifies significant population centers within an 80-km radius [50 mi] {3.2 km [2 mi] for research and development operations} from the approximate center of the projected activities.

(3)   A map of suitable scale is provided, centered on the proposed ISL facility, marked with concentric circles at 1, 2, 3, 4, 5, 10, 20, 30, 40, 50, 60, 70, and 80 km divided into 22½-degree sectors centered on one of the 16 compass points. A table keyed to this map showing separate and cumulative population totals for each sector and annular ring is provided. The distance to the nearest residence is noted for each sector.

(4)   Descriptions of significant population and visitor statistics of neighboring schools, plants, hospitals, sports facilities, residential areas, parks, and forests within 3.2 km [2 mi] of the proposed *in situ* leach facility, based on generally accepted sources such as the U.S. Census Bureau, and State and local agencies, are provided, with identification of data sources.

(5)   Projections are included of population, visitor, and food production data over the expected life of the *in situ* leach facility (typically tens of years).

(6)   Descriptions of the methodology and sources used to develop projections are provided.

The food production data are acceptable if data (kg/yr) for vegetables, meat, and milk, based on generally accepted sources such as the U.S. Department of Agriculture, Farm Bureau, and state and local agriculture services, are provided, with identification of data sources.

2.3.4 Evaluation Findings

If the staff review as described in this section results in the acceptance of the population distribution and food production data, the following conclusions may be presented in the technical evaluation report and in the environmental assessment.

NRC has completed its review of the site characterization information concerned with population distribution and food production near the _____ *in situ* leach facility. This review included an evaluation using the review procedures in standard review plan Section 2.3.2 and acceptance criteria outlined in standard review plan Section 2.3.3.

The applicant has acceptably described the population distribution using population data from generally accepted sources. A map showing the location of significant population centers, within an 80-km radius [50 mi] of the approximate center of proposed operations, is provided. A table and accompanying map providing population in pie-shaped wedges, centered on each of the 16 compass points, is included. Nearest residence distances are noted for each sector. The applicant has provided acceptable information on minority and low-income populations, schools, industrial facilities, sports facilities, residential areas, parks, and forests within 3.2 km [2 mi] of the proposed *in situ* leach facility. Food production data (e.g., vegetables, meat, milk) have been described and keyed on a map. Based on a description of the methodology and sources, all the data have been appropriately projected for the proposed life of the *in situ* leach facility.

Based on the information provided in the application, and the detailed review conducted of the characterization of population distribution and food production for the _____ *in situ* leach facility, the staff concludes that the information is acceptable and is in compliance with 10 CFR 51.45, which requires a description of the affected environment containing sufficient data to aid the Commission in its conduct of an independent analysis.

2.3.5   References

NRC. Regulatory Guide 3.46, "Standard Format and Content of License Applications, Including Environmental Reports, for *In Situ* Uranium Solution Mining." Washington, DC: NRC, Office of Standards Development. 1982.

————. NUREG–1748, "Environmental Review Guidance for Licensing Actions Associated with NMSS Programs." Washington, DC: NRC. 2001.

2.4   Historic, Scenic, and Cultural Resources

2.4.1   Areas of Review

The staff shall review discussions of the historic, cultural, and scenic resources, if any, within the area of potential effect. Historic properties include districts, sites, buildings, structures, or objects of historical, archaeological, architectural, or traditional cultural significance. Specific attention should be directed to properties included in or eligible for inclusion in the National Register of Historic Places (the National Register) and properties registered as National Natural Landmarks.

The staff should review identifications of those properties included in, or eligible for, inclusion in the National Register of Historic Places, located within the area of the proposed project, and should review evidence of contact with the appropriate state historic preservation officer, including a copy of any state historic preservation officer comments concerning the effect of the facility on historic, scenic, and cultural resources.

The review should include information on whether new roads, pipelines, or utilities for the proposed activity will pass through or near any area or location of known historic, scenic, or cultural significance.

Site Characterization

2.4.2 Review Procedures

The staff should determine that the applicant has used the appropriate databases and records to identify historic, scenic, and cultural resources that are found within the study region. The staff should determine that the locations and descriptions of the features are sufficient to allow an evaluation of the likely impacts of the proposed facilities on these resources. Of particular interest are features included in, or eligible for inclusion in, the National Register and National Natural Landmarks. Means to consider and treat such data are discussed in several National Park Service guidelines (e.g., National Park Service, 1973, 1990, 1995). The reviewer should verify that data presented support the of estimates of long-term costs in terms of the likely impacts on the aesthetic or recreational values of such landmarks. It is important that the application document evidence of contact with knowledgeable sources when no historic, scenic, or cultural resources are identified by the applicant within the study area. The reviewer should examine the likely impact of new roads, pipelines, or other utilities on areas and locations of known historic, scenic, or cultural significance [White House, 2000 (Executive Order 13175)].

The reviewer should also confer with the state historic preservation officer as required by 36 CFR Part 800. As specified in Part 800, the state historic preservation officer can enter into a memorandum of understanding to assume the function of the Advisory Council on Historic Preservation. In these situations, consistent with 36 CFR 800.7(b)(1), NRC can comply with the state review process in lieu of the Advisory Council on Historic Preservation regulations. If such a memorandum of understanding is not in place, the staff must consult with the state historic preservation officer and other interested parties. If adverse effects are found, and the Advisory Council on Historic Preservation does not participate, the NRC may enter into a memorandum of agreement with the State Historic Preservation Officer as specified in 36 CFR 800.6(b)(1). The NRC must submit a copy of the executed memorandum of agreement, along with the documentation specified in 36 CFR 800.11(f) to the Advisory Council on Historic Preservation prior to approving the undertaking in order to meet the requirements of Section 106 of the National Historic Preservation Act. If adverse effects are found, and the Advisory Council on Historic Preservation does not participate, the NRC should follow the requirements of 36 CFR 800.6(b)(2).

For license renewals and amendment applications, Appendix A to this standard review plan provides guidance for examining facility operations and the approach that should be used in evaluating amendments and renewal applications.

2.4.3 Acceptance Criteria

The characterization of regional historic, scenic, and cultural resources is acceptable if it meets the following criteria:

(1)   A listing for all properties included in, or eligible for inclusion in, the National Register including National Natural Landmarks is provided.

(2)   A map is included showing all identified National Register Properties and National Natural Landmarks with respect to the location of facilities such as buildings, new roads, well fields, pipelines, surface impoundments, and utilities that might affect these areas.

        A license condition will be placed in the license prohibiting work if any previously unknown cultural artifacts are found.

(3)   Discussions are incorporated of the treatment of areas of historic, scenic, and cultural significance that follow guidance equivalent to that provided by the National Park Service Preparation of Environmental Statements: Guidelines for Discussion of Cultural (Historic, Archeological, Architectural) Resources (National Park Service, 1973). Where appropriate, tribal authorities have been consulted on the likely impacts on Native American cultural resources (White House, 2000). For a consideration of environmental justice, see Section 7.6.1.3, Acceptance Criterion (3) and NUREG–1748 (NRC, 2001).

(4)   If delegated by NRC, the applicant provides evidence of contact with the appropriate state historic preservation officer and tribal authorities. This evidence includes a copy of comments of the state historic preservation officer and tribal authority concerning the effects of the proposed facility on historic, archeological, architectural, and cultural resources.

(5)   If delegated by NRC, the applicant presents a memorandum of agreement among the state historic preservation officer, tribal authorities, and other interested parties regarding their satisfaction with regard to the protection of historic, archeological, architectural, and cultural resources during site construction and operations.

(6)   A letter from the state historic preservation officer has been obtained that discusses any issues associated with sites in, or eligible for inclusion in, the National Register, National Natural Landmarks, or other cultural properties that may be affected by the *in situ* leach operations.

(7)   The aesthetic and scenic quality of the site is rated in accordance with U.S. Bureau of Land Management 8400—Visual Resource Management (U.S. Bureau of Land Management, 2001).

        If the rating is below 19 (scale of 0 to 33), no special management is required. If the rating is 19 or above, the application provides a management plan for minimizing the impact of the proposed facility.

2.4.4   Evaluation Findings

If the staff review as described in this section results in the acceptance of the characterization of the historic, scenic, and cultural resources the following conclusions may be presented in the environmental assessment.

NRC has completed its review of the site characterization information concerned with regional historic, scenic, and cultural resources near the _____ *in situ* leach facility. This review included an evaluation using the review procedures in standard review plan Section 2.4.2 and acceptance criteria outlined in standard review plan Section 2.4.3.

The licensee has acceptably described the historic, scenic, and cultural resources. A listing of all nearby areas and properties included or eligible for inclusion in the National Register or National Natural Landmarks is provided. A map showing all historic landmarks and places with respect to *in situ* leach facilities is included. A record of the investigation of places and properties with historic, scenic, and cultural significance, which follows guidance equivalent to that of the National Park Service, is provided. Contact with local tribal authorities, where appropriate, is acceptably documented. A letter from the state historic preservation officer addressing any issues related to the properties that might be affected by the *in situ* leach facilities is included. The applicant has acceptably demonstrated that the state historic preservation officer and tribal authorities agree with the planned protection from or determination of lack of conflict with *in situ* leach facilities and activities and with any places of importance to the state, federal, or tribal authorities. The applicant has acceptably rated the aesthetic and scenic quality of the site in accordance with the U.S. Bureau of Land Management Visual Resource Inventory and Evaluation System.

Based on the information provided in the application, and the detailed review conducted of the characterization of regional historic, archeological, architectural, scenic, cultural, and natural landmarks near the _____ *in situ* leach facility, the staff concludes that the information is acceptable and is in compliance with 10 CFR 51.45, which requires a description of the affected environment containing sufficient data to aid the Commission in its conduct of an independent analysis.

2.4.5 References

National Park Service. "How to Apply the National Register Criteria for Evaluation." National Park Service Bulletin No. 15. Washington, DC: National Park Service, U.S. Department of the Interior. 1995.

———. "Guidelines for Evaluating and Documenting Traditional Cultural Properties. National Register Bulletin No. 38. Washington, DC: National Park Service, U.S. Department of the Interior. 1990.

———. "Preparation of Environmental Statements: Guidelines for Discussion of Cultural (Historic, Archeological, Architectural) Resources." Washington, DC: National Park Service. 1973.

NRC. NUREG–1748, "Environmental Review Guidance for Licensing Actions Associated with NMSS Programs." Washington, DC: NRC. 2001.

U.S. Bureau of Land Management. "Visual Resource Management." U.S. Bureau of Land Management Manual—8400. Washington, DC: U.S. Department of the Interior. http://lm0005.blm.gov/nstc/rrm/8400.html. 2000.

White House. "Consultation and Coordination with Indian Tribal Governments." Executive Order 13175. *Federal Register.* Vol. 65. pp. 67249–67252. 2000.

2.5 Meteorology

2.5.1 Areas of Review

The staff should review descriptions of the atmospheric diffusion characteristics of the site and its surrounding area based on data collected onsite or at nearby meteorological stations. The data to be reviewed include

(1) National Weather Service station data, including locations of all National Weather Service stations within an 80-km [50-mi] radius; and available joint frequency distribution data by wind direction, wind speed, stability class, period of record, and height of data measurement

(2) On-site meteorological data, including locations and heights of instrumentation, descriptions of instrumentation, and joint frequency distribution data, if National Weather Service data representative of the site are not available

(3) Miscellaneous data, including annual average mixing layer heights, a description of the regional climatology, and total precipitation and evaporation, by month

The staff should also review a discussion of the general climatology including existing air quality, the relationship of the regional meteorological data to the local data, the meteorological impact of the local terrain and large lakes and other bodies of water, and the occurrence of severe weather in the area and its effects. This review should also include data on averages of temperature and humidity.

2.5.2 Review Procedures

The staff should determine whether the application includes sufficient local and regional-scale meteorological information to support estimates of airborne radionuclide transport from the proposed *in situ* leach facility to the surrounding area and for determination of airborne pathway inputs to risk assessment models. This information may include National Weather Service data, on-site monitoring data, or data from local meteorological stations, and any maps or tables that describe meteorological conditions at the site and surrounding area. Section 2.5 of the Standard Format and Content of License Applications, Including Environmental Reports (NRC, 1982) contains a list of acceptable meteorological data requirements.

For license renewals and amendment applications, Appendix A to this standard review plan provides guidance for examining facility operations and the approach that should be used in evaluating amendments and renewal applications.

Site Characterization

2.5.3 Acceptance Criteria

The characterization of the site meteorology is acceptable if it meets the following criteria:

(1)   A description of the general climate of the region and local meteorological conditions is provided, based on appropriate data from National Weather Service, military, or other stations recognized as standard installations.

These data include precipitation, evaporation, and joint-frequency distribution data by wind direction, wind speed, stability class, period of record, and height of data measurement. The average inversion height should also be identified. Data should also be provided on diurnal and monthly averages of temperature and humidity. The locations of all stations used in the data analysis and the height of the data measurement should be included. Data periods should be defined by month and year and cover a sufficient time period to constrain long-term trends and support atmospheric dispersion modeling.

Data from local meteorological weather stations supplemented, if necessary, by data from an on-site monitoring program, are provided.

A minimum of one full year of joint frequency data presented with a joint data recovery of 90 percent or more is provided.

The on-site program should be designed in accordance with Regulatory Guide 3.63, "Onsite Meteorological Measurement Program for Uranium Recovery Facilities—Data Acquisition and Reporting" (NRC, 1988).

(2)   Consideration of relationships between regional weather patterns and local meteorological conditions based on weather station data and the on-site monitoring program, if necessary, is included. The impacts of terrain and nearby bodies of water on local meteorology are assessed, and the occurrence of locally severe weather is described and its impact considered.

Information on anticipated air quality impacts from non-radiological sources, such as vehicle emissions and dust from well field activities, is provided for assessing cumulative impacts.

(3)   The meteorological data used for assessing impacts are substantiated as being representative of expected long-term conditions at and near the site.

(4)   The application contains a description of existing air quality.

The applicant must demonstrate that the radiological and non-radiological air quality impacts caused by *in situ* leach facilities are virtually indistinguishable from background, or information on the likelihood of air pollution is based on U.S. Environmental Protection Agency (EPA) studies. Affected counties within 80 km [50 mi] of the facility are classified according to the National Ambient Air Quality

Standards as being in attainment (below National Ambient Air Quality Standards) or nonattainment (above National Ambient Air Quality Standards status.

(5)   The sources of all meteorological and air quality data are documented in open file reports or other published documents. If data have been generated by the applicant the data documentation should include a description of the investigations and data reduction techniques.

2.5.4   Evaluation Findings

If the staff review as described in this section results in the acceptance of the meteorology, the following conclusions may be presented in the technical evaluation report and in the environmental assessment.

NRC has completed its review of the site characterization information concerned with meteorology at the _____ in situ leach facility. This review included an evaluation using the review procedures in standard review plan Section 2.5.2 and acceptance criteria outlined in standard review plan Section 2.5.3.

The licensee has acceptably described the site meteorology by providing data from National Weather Service military, or other stations recognized as standard installations located within 80 km [50 mi] of the site, including available joint frequency distribution data on (i) wind direction and speed, (ii) stability class, (iii) period of record, (iv) height of data measurement, and (v) average inversion height. The data cover a sufficient time period to constrain long-term trends and support atmospheric dispersion modeling. The applicant has provided acceptable on-site meteorological data, if necessary, including (i) descriptions of instruments, (ii) locations and heights of instruments, and (iii) joint frequency distributions. The joint-frequency data presented are for a minimum of 1 year, with a joint data recovery of 90 percent or more. Additional data on (i) annual average mixing layer heights, (ii) a description of the regional climate, and (iii) total precipitation and evaporation by month have been provided. The applicant has noted any effect of nearby water bodies or terrain on meteorologic measurements. The applicant has acceptably demonstrated that meteorologic data used for assessing environmental impacts are representative of long-term meteorologic conditions at the site. The applicant report on the existing air quality at the site and nearby is acceptable.

Based on the information provided in the application, and the detailed review conducted of the characterization of meteorology at the _____ in situ leach facility, the staff concludes that the information is acceptable to allow evaluation of the spread of airborne contamination at the site and development of conceptual and numerical models, and is in compliance with 10 CFR 51.45, which requires a description of the affected environment containing sufficient data to aid the Commission in its conduct of an independent analysis. The characterization also meets the requirements of 10 CFR Part 40, Appendix A, Criterion 7, which requires pre-operational and operational monitoring programs.

2.5.5 References

NRC. Regulatory Guide 3.63, "Onsite Meteorological Measurement Program for Uranium Recovery Facilities—Data Acquisition and Reporting." Washington, DC: NRC, Office of Standards Development. 1988.

————. Regulatory Guide 3.46, "Standard Format and Content of License Applications, Including Environmental Reports, for *In Situ* Uranium Solution Mining." Washington, DC: NRC, Office of Standards Development. 1982.

2.6   Geology and Seismology

2.6.1   Areas of Review

The reviewer should examine information on the geologic aspects of the site acquired through standard geologic analyses, including a survey of pertinent literature and field investigations. This information should include regional seismicity and seismic history, local stratigraphy, petrology or lithology of rock units, tectonic features (faulting, folding, fracturing), and the continuity of the geologic strata at the site and in nearby regions.

Geologic, structural, and stratigraphic maps and cross sections, including representative core and geophysical well-log data of the site and its environs, should be reviewed. An isopach map of the intended zone of injection or production and associated confining beds should be evaluated. All conclusions regarding the lateral continuity and vertical thickness of the mineralized zone(s), surrounding lithologic units, and confining zones, as based on lithologic logs from core and drill cuttings, geophysical data, remote-sensing measurements, and the results of other appropriate investigations should be reviewed. Some of the applicant's supporting information for this review area might be included in the documents submitted to satisfy the hydrology review area (Section 2.7).

The staff should review the information presented on any economically important minerals and energy-related deposits in addition to the uranium mineralization, including the likely consequences of any production of such related deposits on the *in situ* leach facility.

Data on the geochemistry of the ore zone and the geologic zones immediately surrounding the mineralized zone that will or could be affected by injected lixiviant should be evaluated. Information on unique minerals (including those that might be affected by fluid movement associated with the proposed project, such as bentonite) or paleontologic deposits of particular scientific interest, should also be reviewed. The staff should examine descriptions of any effects that planned operations at the site might have on the future availability of other mineral resources.

2.6.2 Review Procedures

The staff should review the application to determine whether a thorough evaluation of the geologic setting for the proposed *in situ* leach activity has been presented along with the basic data supporting all conclusions. In addition to a description of the basic geology, both at the surface and at the depths of interest, the establishment of the continuity of the geologic strata at the site should be reviewed for applicability, correctness, inclusivity, and likely ability of the strata to isolate *in situ* leach fluids. The reviewer should particularly focus attention on fractures or faults, permeable stratigraphic units, and lateral facies changes that might preclude the applicant-identified geologic barriers to fluid migration from performing adequately.

The reviewer should determine that the application contains accurate geologic maps, isopach maps of the mineralized strata and of the confining layers, geologic cross sections at places critical to a thorough understanding of the selected site, descriptions of representative supporting core samples, geophysical and lithologic logs, and other data required for a thorough understanding of the pertinent geology. The reviewer should determine that regional stratigraphic and geologic information is discussed in sufficient detail to give clear perspective and orientation to the site-specific material presented. The discussion of regional geology and stratigraphy should be assessed to determine if it is adequately referenced and is illustrated by regional surface and subsurface geologic maps, stratigraphic columns, and cross sections. Seismic information should be evaluated to assess its suitability for evaluating seismic hazard for the proposed facility.

The staff may also perform an independent analysis of the data provided to assess whether reasonable and conservative alternative interpretations are indicated.

For license renewals and amendment applications, Appendix A to this standard review plan provides guidance for examining facility operations and the approach that should be used in evaluating amendments and renewal applications.

2.6.3 Acceptance Criteria

The characterizations of the site geology and seismology are acceptable if they meet the following criteria:

(1)   The application includes a description of the local and regional stratigraphy based on techniques such as

    (a)   Surface sampling and descriptions

    (b)   Cuttings and core logging reports

    (c)   Wireline geophysical logs, such as electrical resistivity, neutron density, and gamma logs

    (d)   Geologic interpretations of surface geology and balanced cross sections

These interpretations may be based either on original work submitted by the applicant, or on an appropriate evaluation of previous work in the region performed by state or federal agencies (e.g., U.S. Geological Survey, U.S. Bureau of Land Reclamation, U.S. Bureau of Mines), universities, mining companies, or oil and gas exploration companies. The interpretations should be accompanied by

(i)   Maps such as geologic, topographic, and isopach maps that show surface and subsurface geology and locations for all wells used in defining the stratigraphy

(ii)   Cross sections through the ore deposit roughly perpendicular and parallel to the principal ore trend

(iii)   Fence diagrams showing stratigraphic correlations among wells

(2)   All maps and cross sections are at sufficient scale and resolution to show clearly the intended geologic information. Maps show the locations of all site explorations such as borings, trenches, seismic lines, piezometer readings, and geologic cross sections.

(3)   In the local stratigraphic section, all mineralized horizons, confining units, and other important units such as drinking water aquifers and deep well injection zones are clearly shown, with their depths from the surface clearly indicated. Isopach maps are prepared showing the variations in thickness of the mineralized zones and the confining units over the proposed mining area.

(4)   A geologic and geochemical description of the mineralized zone and the geologic units immediately surrounding the mineralized zone is provided.

(5)   An inventory of economically significant mineral and energy-related deposits, in addition to the uranium mineralization, is provided. Locations of all known wells, surface and underground mine workings, and surface impoundments that may have an effect on the proposed operations are provided.

These items should be located on a map of sufficient scale and clarity to identify their relationship to the proposed facility. For existing wells, the depth should be shown, if possible. To allow evaluation of connections between the mineralized zone and underground sources of drinking water, plugging and abandonment records provided from state, federal, and local sources, as appropriate, should be provided. The applicant should provide evidence that action has been undertaken to properly plug and abandon all wells that cannot be documented in this manner.

(6)   A description of the local and regional geologic structure, including folds and faults, is provided.

Folds and faults can be shown on the geologic maps used to describe the stratigraphy. Major and minor faults traversing the proposed site should be evaluated for the likely consequences of any future effects of faulting on the uranium production activities and on the ability of the strata to contain lixiviant should fault motion occur. Geologic structures that are preferential pathways or barriers to fluid flow must be described and the basis for likely effects on flow given.

(7)   A discussion of the seismicity and the seismic history of the region is included.

Historical seismicity data should be summarized on a regional earthquake epicenter map, including magnitude, location, and date of all known seismic events. Where possible, seismic events should be associated with the tectonic features described in the geologic structures.

(8)   A generalized stratigraphic column, including the thicknesses of rock units, representation of lithologies, and definition of the mineralized horizon, is presented.

(9)   The sources of all geological and seismological data are documented in U.S. Geological Survey open files or other published documents. If data have been generated by the applicant, the documentation should include a description of the investigations and data reduction techniques.

(10)   Maps have designation of scale, orientation (e.g., North arrow), and geographic coordinates.

(11)   Short-term seismic stability has been demonstrated for the *in situ* leach facility in accordance with Regulatory Guide 3.11, "Design, Construction, and Inspection of Embankment Retention Systems for Uranium Mills," Section 2.6 (NRC, 1977).

(12)   A general description of the site soils and their properties has been provided to support an evaluation of the environmental effects of construction and operation on erosion.

(13)   A detailed description of soils and their properties has been provided for any areas where land application of water is anticipated to support an assessment of the impacts.

2.6.4 Evaluation Findings

If the staff review as described in this section results in the acceptance of the characterization of the geology and seismology, the following conclusions may be presented in the technical evaluation report and in the environmental assessment.

NRC has completed its review of the site characterization information concerned with geology and seismology at the _____ *in situ* leach facility. This review included an evaluation using the review procedures in standard review plan Section 2.6.2 and acceptance criteria outlined in standard review plan Section 2.6.3.

The licensee has acceptably described the geology and seismology by providing (i) a description of the local and regional stratigraphy; (ii) geologic, topographic, and isopach maps at acceptable scales showing surface and subsurface features and locations of all wells and site explorations used in defining stratigraphy; (iii) a geologic and geochemical description of the mineralized zone and the geologic units adjacent to the mineralized zone; (iv) an inventory of nearby economically significant minerals and energy-related deposits; (v) a description of the local and regional geologic structure; (vi) a discussion of the seismicity and seismic history of the region; (vii) a generalized stratigraphic column that includes thickness of rock units, representation of lithologies, and definition of mineralized horizon; and (viii) a description and map of the soils.

Based on the information provided in the application, and the detailed review conducted of the characterization of the geology and seismology at the _____ *in situ* leach facility, the staff concludes that the information is acceptable to allow evaluation of the geologic and seismologic characteristics of the site, supports associated conceptual and numerical models, and is in compliance with 10 CFR 40.31(f), which requires inclusion of an environmental report in the application, and 10 CFR 51.45, which requires a description of the affected environment containing sufficient data to aid the Commission in its conduct of an independent analysis. The characterization is sufficient to meet the requirements of 10 CFR Part 40, Appendix A, Criteria 4(e), which requires locations away from faults capable of causing impoundment failure and 5G(2), which requires adequate descriptions of the characteristics of the underlying soils and geologic formations.

2.6.5 Reference

NRC. Regulatory Guide 3.11, "Design, Construction, and Inspection of Embankment Retention Systems for Uranium Mills." Washington, DC: NRC, Office of Standards Development. 1977.

2.7 Hydrology

2.7.1 Areas of Review

Characterization of the hydrology at *in situ* leach uranium extraction facilities must be sufficient to establish potential effects of *in situ* leach operations on the adjacent surface-water and ground-water resources and the potential effects of surface-water flooding on the *in situ* leach facility. The areas of review include:

(1)   Descriptions of surface-water features in the site area including type, size, pertinent hydrological or morphological characteristics, and proximity to *in situ* leach processing plants, well fields, evaporation ponds, or other facilities that might be negatively affected by surface erosion or flooding.

(2)   Assessment of the potential for erosion or flooding that may require special design features or mitigation measures to be implemented.

(3)   A description of site hydrogeology, including (i) identification of aquifer and aquitard formations that may affect or be affected by the *in situ* leach operations; (ii) a description of aquifer properties, including material type, formation thickness, effective porosity, hydraulic conductivity, and hydraulic gradient; (iii) estimated thickness and lateral extent

of aquitards, and other information relative to the control and prevention of excursions; and (iv) data to support conclusions concerning the local ground-water flow system, based on well borings, core samples, water-level measurements, pumping tests, laboratory tests, soil surveys, and other methods

(4)   Assessment of available ground-water resources and ground-water quality within the proposed permit boundaries and adjacent properties, including quantitative description of the chemical and radiological characteristics of the ground water and potential changes in water quality caused by operations

(5)   An assessment of typical seasonal ranges and averages and the historical extremes for levels of surface-water bodies and aquifers

(6)   Information on past, current, and anticipated future water use, including descriptions of local ground-water well locations, type of use, amounts used, and screened intervals

In conducting these evaluations, the reviewer shall consider the technical evaluations conducted by a state or another federal agency with authorities overlapping those of the NRC. Ground-water compliance and protection reviews are the primary technical areas impacted by overlapping authorities. The desired outcome is to identify any areas where duplicative NRC reviews may be reduced or eliminated. The NRC staff must make the necessary evaluations of compliance with applicable regulations for licensing the facility. However, the reviewer may, as appropriate, rely on the applicant's responses to inquiries made by a state or another federal agency to support the NRC evaluation of compliance. The reviewer should make every effort to coordinate the NRC technical review with the state or other federal agency with overlapping authority to avoid unnecessary duplication of effort.

2.7.2   Review Procedures

At a minimum, the reviewer should evaluate whether the applicant has developed an acceptable conceptual model of the site hydrology and whether the conceptual model is adequately supported by the data presented in the site characterization. To this end, the reviewer should:

(1)   Review surface water data, including maps that identify nearby lakes, rivers, surface drainage areas, or other surface-water bodies; stream flow data; and the applicant's assessment of the likely consequences of surface-water contamination from *in situ* leach operations. Verify that the applicant has generally characterized perennial surface-water bodies, such that an assessment of impacts from operations can be made.

(2)   Evaluate the applicant's assessment of the potential for erosion or flooding. If surface water or erosion modeling is used by the applicant, verify that acceptable models and input parameters have been used in the flood analyses and that the resulting flood forces have been acceptably accommodated in the design of surface impounds. Regardless of whether modeling is used, ensure that the evaluation of flooding and

erosion potential is consistent with available geomorphological, and topographic data or analysis of paleodischarge information.

(3)   Evaluate the site hydrogeologic conceptual model for ground-water flow in potentially affected aquifers. Review available data from well logs and hydrologic tests and measurements to obtain confidence that sufficient data have been collected and that the data support the applicant's hydrologic conceptual model for ground-water flow within and around the permit boundary. The applicant's interpretation of ground-water hydraulic gradients (used to infer flow direction), horizontal hydraulic conductivity, and the thickness, areal extent, and vertical hydraulic conductivity of confining formations should be evaluated. Examine pumping tests, analyses, and/or other measurement techniques used to determine the hydrologic properties of the local aquifers and aquitards that affect or may be affected by the proposed *in situ* leach activities. Also examine pumping tests that are used to investigate vertical confinement or hydraulic isolation between the ore production zone and upper and lower aquifers.

(4)   Evaluate the applicant's assessment of water quality of potentially affected ground-water resources. This information will provide the basis for evaluating potential effects of *in situ* leach extraction on the quality of local ground-water resources. Verify that a sufficient number of baseline ground-water samples are collected to provide meaningful statistics, that samples are spaced in time sufficiently to capture temporal variations, and that the chemical constituents and water quality parameters evaluated are sufficient to establish pre-operational water quality, including classes of use.

(5)   Review the applicant's assessment of seasonal and, if data are available, the historical variability for levels of surface-water bodies and water levels or potentiometric heads in aquifers and ensure that sufficient time intervals have elapsed between measurements to allow assessment of seasonal variability.

(6)   Verify that the applicant has provided information on past, current, and anticipated future water uses, including descriptions of local ground-water well locations, type of use, amounts used, and screened intervals.

In conducting an evaluation of ground-water activities, the reviewer should follow the reviews conducted by the state. Where appropriate, the evaluation should not duplicate state regulatory efforts. Although NRC must make its own independent findings, reviewers need not duplicate questions if a state or other federal regulatory agency has already addressed the issue. If the applicant response to questions from a state or other federal agency is submitted to NRC so that it becomes part of the license application to NRC, then the reviewer can use the information to prepare the technical evaluation report on ground-water issues.

2.7.3   Acceptance Criteria

The hydrologic characterization should establish a hydrologic conceptual model for the *in situ* leach site and surrounding region. The conceptual model provides a framework for the applicant to make decisions on the optimal methods for extracting uranium from the mineralized zones, and to minimize environmental and safety concerns caused by *in situ* leach operations. Hydrologic characterizations that accomplish this objective are considered acceptable.

The characterization of the site hydrology is acceptable if it meets the following criteria:

(1)   The applicant has characterized surface-water bodies and drainages within the licensed area and affected surroundings. Maps provided in the application identify the location, size, shape, hydrologic characteristics, and uses of surface-water bodies near the proposed site, including likely surface drainage areas near the proposed facilities. An acceptable application should also identify the zones of interchange between surface water and ground water.

(2)   The applicant has provided an assessment of the potential for flooding and erosion that could affect the *in situ* leach processing facilities or surface impoundments. The staff recognizes that the flooding and erosion protection design of impoundments for *in situ* facilities may be relatively simple. This is true when impoundments are located near or on a drainage divide and little or no diversion of runoff is necessary to protect the impoundment side slopes from erosion. In such cases, it will be easy to demonstrate that no erosion to the slopes will occur. In flood-prone areas, however, it may be necessary to conduct surface water and erosion modeling. Information regarding acceptable models may be found in NUREG–1623 (NRC, 1999). The reviewer should recognize, however, that the staff guidance (NRC, 1999) was prepared for use in evaluating a 1,000-year design life for large tailings impoundments, whereas the design life of the surface impoundments at *in situ* leach facilities is on the order of tens of years.

(3)   The applicant has described the local and regional hydraulic gradient and hydrostratigraphy. The applicant has shown that subsurface water level measurements were collected by acceptable methods, such as American Society for Testing and Materials D4750 (American Society for Testing and Materials, 2001). Potentiometric maps are the recommended means for presenting hydraulic gradient data. These maps should include two levels of detail: regional and local. The regional map should represent the mineralized zone aquifer and should encompass the likely consequences on any affected highly populated areas. The local (site-scale) map should encompass the entire licensed area. If overlying and underlying aquifers exist, local-scale potentiometric or water surface elevation maps of these aquifers should also be included. These maps should clearly show the locations, depths, and screened intervals of the wells used to determine the potentiometric surface elevations. Alternatively, this information can be provided in separate maps and/or tables. The appropriate contour interval will vary from site to site; however, contour intervals should be sufficient to clearly show the ground-water flow direction in the ore zone and in the overlying and underlying aquifers. The number of piezometer elevation measurements used to construct each map should be sufficient to determine the direction of ground-water flow in the mineralized zone(s) and the overlying aquifer. To construct a regional potentiometric map, a reasonable effort should be made to consider as many existing wells as possible.

Site Characterization

Hydrogeologic cross sections are recommended for illustrating the interpreted hydrostratigraphy. These cross sections should be constructed for the area within the license boundary. For very large or irregularly shaped well field areas, more than one cross section may be necessary. Cross sections must be based on borehole data collected during well installation or exploratory drilling. All significant borehole data should be included in an appendix. Staff should verify that, an adequate number of boreholes is used to support the assertion of hydrogeologic unit continuity, if shown as such in the cross sections.

The applicant should describe all hydraulic parameters used to determine expected operational and restoration performance. Aquifer and aquitard hydraulic properties may be determined using aquifer pumping tests for parameters such as hydraulic conductivity, transmissivity, and specific storage. Any of a number of commonly used aquifer pumping tests may be used including single-well drawdown and recovery tests, drawdown versus time in a single observation well, and drawdown versus distance pumping tests using multiple observation wells. The methods or standards used to analyze pumping test data should be described and referenced: acceptable methods of analysis include use of curve fitting techniques for drawdown or recovery curves that are referenced to peer-reviewed journal publications, texts, or American Society for Testing and Materials Standards. It is important for the reviewer to ensure that where fitted curves deviate from measured drawdown, the applicant explains the probable cause of the deviation (e.g., leaky aquitards, delayed yield effects, boundary effects, etc.). For estimates of porosity, it is acceptable to use laboratory analysis of core samples, borehole geophysical methods, and analysis of the barometric efficiency of the aquifer (e.g., Lohman, 1979). The applicant should distinguish between total porosity estimated from borehole geophysical methods and effective porosity that determines transport of chemical constituents.

(4)   Reasonably comprehensive chemical and radiochemical analyses of water samples, obtained within and at locations away from the mineralized zone(s), have been made to determine pre-operational baseline conditions. Baseline water quality should be determined for the mineralized and surrounding aquifers. These data should include water quality parameters that are expected to increase in concentration as a result of *in situ* leach activities and that are of concern to the water use of the aquifer (i.e., drinking water, etc.). The applicant should show that water samples were collected by acceptable sampling procedures, such as American Society for Testing and Materials D4448 (American Society for Testing and Materials, 1992).

For example, *in situ* leach operations are not expected to mobilize aluminum, and unless an ammonia-based lixiviant is used, ammonia concentrations in the ground water should not increase as a result of *in situ* leach operations. Therefore, little is gained by sampling these parameters. Studies have shown that thorium-230 is mobilized by bicarbonate-laden leaching solutions. However, studies have also shown that after restoration, thorium in the ground water will not remain in solution because the chemistry of thorium causes it to precipitate and chemically react with the rock matrix (Hem, 1970). As a result of its low solubility in natural waters, thorium is found in only trace concentrations. Additionally, chemical tests for thorium are expensive, and are not commonly included in water analyses at *in situ* leach facilities.

Table 2.7.3-1. Typical Baseline Water Quality Indicators to be Determined During Pre-operational Data Collection		
A. Trace and Minor Elements		
Arsenic	Iron	Selenium
Barium	Lead	Silver
Boron	Manganese	Uranium
Cadmium	Mercury	Vanadium
Chromium	Molybdenum	Zinc
Copper	Nickel	
Fluoride	Radium-226@	
B. Common Constituents		
Alkalinity	Chloride	Sodium
Bicarbonate	Magnesium	Sulfate
Calcium	Nitrate	
Carbonate	Potassium	
C. Physical Indicators		
Specific Conductivity*		Total Dissolved Solids#
pH*		
D. Radiological Parameters		
Gross Alpha†	Gross Beta	
*Field and Laboratory determination. #Laboratory only. †Excluding radon, radium, and uranium. @ If site initial sampling indicates the presence of Th-232 then Ra-228 should be considered in the base line sampling or an alternative may be proposed.		

Site Characterization

The applicant should identify the list of constituents to be sampled for baseline concentrations. The list of constituents in Table 2.7.3-1 is accepted by the NRC for *in situ* leach facilities. Alternatively, applicants may propose a list of constituents that is tailored to a particular location. In such cases, sufficient technical bases must be provided for the selected constituent list.

At least four sets of samples, spaced sufficiently in time to indicate seasonal variability, should be collected and analyzed for each listed constituent for determining baseline water quality conditions. Some samples should be split and sent to different laboratories as part of a quality assurance program. Sets of samples should be taken with a minimum of a week or two between sampling to provide an indication of how the water quality of the aquifers changes with time. The applicant should document any variability in the ground-water flow rates or recharge that are observed in the collected data. Additional sampling to establish the natural cyclical fluctuations of the water quality is necessary if natural ground-water flow rates and recharge conditions vary considerably. Where perennial surface-water sources are present, surface-water quality measurements should be taken on a seasonal basis for a minimum of 1 year before implementation of *in situ* leach operations. Surface-water samples can be obtained by grab sampling and should be taken at the same location each time. The average water quality for each aquifer zone and the range of each indicator in the zone have been tabulated and evaluated. If zones of distinct water quality characteristics are identified, they are delineated and referenced on a topographic map. For example, since uranium rollfront deposits are formed at the interface between chemically oxidizing and reducing environments, water quality characteristics may differ significantly across the rollfront.

(5) The applicant has provided an assessment of seasonal and the historical variability for potentiometric heads and hydraulic gradients in aquifers and water levels of surface-water bodies. This assessment should include water levels or water potentials measurements over at least 1 year and collected periodically to represent any seasonal variability.

(6) The applicant has provided information on past, current, and anticipated future water use, including descriptions of local ground-water well locations, type of use, amounts used, and screened intervals. This information must be sufficient to evaluate potential risks to ground-water or surface-water users in the vicinity of the *in situ* leach facility.

For license renewals and amendment applications, most or all of the preceding acceptance criteria may previously have been met. Appendix A to this standard review plan provides guidance for examining facility operations and the approach that should be used in evaluating amendments and renewal applications.

2.7.4 Evaluation Findings

If the staff's review as described in this section results in the acceptance of the site hydrology, the following conclusions may be presented in the technical evaluation report and in the environmental assessment.

NRC has completed its review of the hydrologic site characterization information for the _____ in situ leach facility. This review included an evaluation using the review procedures in standard review plan Section 2.7.2 and acceptance criteria outlined in standard review plan Section 2.7.3.

The licensee has acceptably described the hydrology by providing (i) estimates of the local and regional hydraulic gradients, using potentiometric surface maps with acceptable contour intervals, including the mineralized aquifer and other overlying or underlying aquifers, and the likely consequences to affected populated areas; (ii) hydrologic cross-sections, based on an appropriate number of boreholes; (iii) acceptable comprehensive chemical and radiochemical analyses of water samples from in and near the mineralized zone(s) that define the pre-operational baseline water quality conditions; (iv) all hydraulic parameters used to determine expected operational and restoration performance; and (v) characterization of surface water in the in situ leach facility and nearby areas, including presentation of such information on maps. Zones of interchange between surface and ground water have been identified. The applicant has provided acceptable erosion protection against the effects of flooding from nearby streams and for drainage and diversion channels, such that the suggested criteria in NUREG–1620 (NRC, 2002) have been followed and that the design meets the requirements of 10 CFR Part 40, Appendix A.

Based on the information provided in the application, and the detailed review conducted of the characterization of the hydrology at the _____ in situ leach facility, the staff concludes that the information is acceptable to allow evaluation of the site and associated conceptual and numerical models and is in compliance with 10 CFR 51.45, which requires a description of the affected environment containing sufficient data to aid the Commission in its conduct of an independent analysis.

2.7.5 References

American Society for Testing and Materials. "Standard Test Method for Determining Subsurface Liquid Levels in a Borehole or Monitoring Well (Observation Well)." Test Method D4750-87. West Conshohcken, Pennsylvania: American Society for Testing and Materials. 2001.

———. "Standard Guide for Sampling Groundwater Monitoring Wells." Guide D4448-85a. West Conshohcken, Pennsylvania: American Society for Testing and Materials. 1992.

Crippen, J.R. and C.D. Bue. "Maximum Floodflows in the Conterminous United States." USGS Water Supply Paper No. 1887. Denver, Colorado: U.S. Geological Survey. 1977.

Hem, J.D. "Study and Interpretation of the Chemical Characteristics of Natural Water." USGS Water Supply Paper 1473. Denver, Colorado: U.S. Geological Survey. 1970.

Lohman, S.W. "Groundwater Hydraulics." USGS Professional Paper 708. Reston, Virginia: U.S. Geological Survey. 1979.

Site Characterization

NRC. NUREG–1620, "Standard Review Plan for the Review of a Reclamation Plan for Mill Tailings Sites Under Title II of the Uranium Mill Tailings Radiation Control Act." Rev. 1. Washington, DC: NRC. 2002.

U.S. Army Corps of Engineers. "Flood Hydrograph Package." HEC–1. Washington, DC: U.S. Army Corps of Engineers, Hydrologic Engineering Center. 1997a.

———. "Water Surface Profiles." HEC–2. Davis, California: Hydrologic Engineering Center. 1997b.

———. "Wave Runup and Wind Setup on Reservoir Embankments." ETL 1110-2-221. 1966.

U.S. Bureau of Reclamation. "Comparison of Estimated Maximum Flood Peaks with Historic Floods." Washington, DC: U.S. Department of the Interior. 1986.

2.8   Ecology

2.8.1   Areas of Review

The staff should review descriptions of the flora and fauna in the vicinity of the licensed area, their habitats, and their distribution. The review should include identification of important species that are (i) threatened or endangered, (ii) commercially or recreationally valuable, (iii) affecting the well-being of some important species within Criterion (i) or (ii), or (iv) critical to the structure and function of the ecological system or a biological indicator of radionuclides or chemical pollutants in the environment.

The review should include the inventory of the majority of the terrestrial and aquatic organisms on or near the site and their relative (qualitative) abundance, the quantitative abundance of the important species, and species that migrate through the area or use it for breeding grounds. The staff should review discussions of the relative importance of the proposed site environs to the total regional area for the living resources (potential or exploited).

For operations involving drying of yellowcake, disposal of waste or generation of hazardous effluents, the staff should examine data on the count and distribution of important domestic fauna, in particular cattle, sheep, and other meat animals that may be involved in the exposure of man to radionuclides. Important game animals should receive similar treatment. A map showing the distribution of the principal plant communities should be reviewed.

The staff should also review the discussion of species-environment relationships, including descriptions of area usage (e.g., habitat, breeding) for important species, life histories of important regional animals and aquatic organisms, normal seasonal population fluctuations and habitat requirements, and identification of food chains and other interspecies relationships, particularly when these contribute to prediction or evaluation of the impact of the facility on the regional biota. The staff should examine any information presented on definable pre-existing environmental stresses from sources such as pollutants, as well as pertinent ecological conditions suggestive of such stresses and the status of ecological succession. As appropriate, the staff should review a list of pertinent published material dealing with the ecology of the region and ecological or biological studies of the site or its environs currently in progress or planned.

2.8.2 Review Procedures

The reviewer should consult with the U.S. Fish and Wildlife Service using procedures in 50 CFR Part 402, "Interagency Cooperation—Endangered Species Act of 1973," as amended. The staff should review the descriptions and inventories of the flora and fauna in the vicinity of the site, including habitats and distribution. The review should include terrestrial and aquatic organisms on or near the site, and their relative (qualitative) abundance should be established. Particular attention should be given to species based on their relative importance to the community. The reviewer should determine that all important species have been identified. Important species should be a part of the larger inventory of species. If important species are determined to be present, the staff should evaluate any likely detrimental effects on the organism by the proposed facility and its operations.

The reviewer should determine that information on the various species is presented in two separate subsections: terrestrial ecology and aquatic ecology. The reviewer should also determine that the discussion of the species-environment relationships includes descriptions of area usage (e.g., habitat, breeding) for important species and discussions of life histories of important regional animals and aquatic organisms, including normal seasonal population fluctuations and their habitat requirements. Food chains and other interspecies relationships should be examined, particularly when these may bear on predictions or evaluations of the impact of the proposed facility on the stability of regional biota. The reviewer should also examine documentation provided for any pre-existing environmental stresses from sources such as pollutants, as well as pertinent ecological indicators suggestive of such stresses. A discussion of the status of ecological succession should be evaluated.

For any operation involving the drying of yellowcake, disposal of waste, or generation of hazardous effluents, the staff should review data on the number and distribution of locally significant domestic flora and fauna, in particular cattle, sheep, commercial fish, and other meat animals, and commercial crops that may be part of the food chain delivering radiation exposure to man. Important game animals should be treated similarly. A map showing the distribution and estimates of numbers of commercially significant species should be examined. Specific review guidance is provided in NUREG–1748 (NRC, 2001).

For license renewals and amendment applications, Appendix A to this standard review plan provides guidance for examining facility operations and the approach that should be used in evaluating amendments and renewal applications.

2.8.3 Acceptance Criteria

The characterization of the site ecology is acceptable if it meets the following criteria:

(1)   Inventories of terrestrial and aquatic species are compiled by the applicant based on reports or databases of state or federal agencies (e.g, U.S. Fish and Wildlife Service, EPA).

Historical sitings of important species, as defined in the Standard Format and Content of License Applications, Including Environmental Reports (NRC, 1982) should be included in the inventory. If such reports do not exist, inventories should be prepared by the applicant based on a radius within which impacts are reasonably expected to occur. Documentation should be provided that inventories were prepared in consultation with appropriate local, state, and federal agencies to confirm the presence or absence of important species (especially threatened or endangered species). Inventories may be based on historical data, but should be updated to within 2 years of the time of application to establish current baselines.

(2)   Inventories of locally significant domestic flora and fauna, in particular cattle, sheep, commercial fish, and other meat-producing animals and commercial crops are based on recent production figures from local, state, and federal agencies (e.g., U.S. Department of Agriculture).

The statistics should cover at least 3 years and have been conducted within 2 years of the date of the application to establish reasonable baselines. Important game animals should be treated similarly. A map showing the distribution and estimates of numbers of commercially significant species should be provided and may be combined with land use maps discussed in Section 2.2 of the standard review plan.

(3)   The applicant has identified any endangered species as listed in 50 CFR Part 17, "Endangered and Threatened Wildlife and Plants."

Any discussion should include nonpermanent inhabitants migrating through the area or using it for breeding grounds. The preservation of habitat, particularly for important species, should be a prime consideration. A map of the principal floral and faunal communities has been provided. Additional information can be found in 50 CFR Parts 401–453.

(4)   The application provides a thorough description of the species-environment relationships for each important species identified within a radius where impacts are reasonably expected to occur. If no important species are identified within this radius, the application should plainly state so, and no additional review is necessary.

The application should take these relationships into account in providing a discussion of any likely detrimental effects that operation of the site may have on the species through changes in habitat, pollution, and aspects of the operations that may place stress on the species-environment relationship. Finally, the application should provide information regarding steps that will be taken to minimize the effect of operating the facility on the species-environment relationship.

(5)   All sources of ecological information are documented in open file reports or other published documents. If data have been generated by the applicant, the documentation should provide a description of the investigations and data reduction techniques.

A list of pertinent published material dealing with the ecology of the region should be included. Any ecological or biological study of the site or its environs either in progress or planned should be described and referenced.

2.8.4   Evaluation Findings

If the staff review as described in this section results in the acceptance of the description of the site ecology, the following conclusions may be presented in the technical evaluation report and in the environmental assessment.

NRC has completed its review of the site characterization information concerned with ecology at the _____ *in situ* leach facility. This review included an evaluation using the review procedures in standard review plan Section 2.8.2 and acceptance criteria outlined in standard review plan Section 2.8.3.

The licensee has described the ecology by providing acceptable (i) inventories of terrestrial and aquatic species, including threatened or endangered species listed in 50 CFR Part 17 (ii) inventories of locally significant domestic flora and fauna (e.g., cattle, sheep, goats), (iii) discussions of important species found within a radius where impacts are reasonably expected to occur and estimations of their current and historical abundance, and (iv) descriptions of the species-environment relationships for any important species.

Based on the information provided in the application and the detailed review conducted of the characterization of the ecology at the _____*in situ* leach facility, the staff concludes that the information is acceptable to allow evaluation of the site ecology and associated conceptual and numerical models and is in compliance with 10 CFR 51.45, which requires a description of the affected environment containing sufficient data to aid the Commission in its conduct of an independent analysis.

2.8.5   References

NRC. NUREG–1748, "Environmental Review Guidance for Licensing Actions Associated with NMSS Programs." Washington, DC: NRC. 2001.

——. Regulatory Guide 3.46, "Standard Format and Content of License Applications, Including Environmental Reports, for *In Situ* Uranium Solution Mining." Washington, DC: NRC, Office of Standards Development. 1982.

2.9   Background Radiological Characteristics

2.9.1   Areas of Review

The reviewer should examine site-specific radiological data provided in the application including the results of measurements of radioactive materials occurring in important species, soil, air, and in surface and ground waters that could be affected by the proposed operations. The reviewer should examine the design of the pre-operational monitoring program, including which radionuclides are analyzed, sampling locations, sample type, sampling frequency, location and density of monitoring stations, and the detection limits.

2.9.2   Review Procedures

The reviewer should examine data from the pre-operational monitoring program with particular attention paid to the design of the monitoring program, the radionuclides monitored, the results, and the detection limits reported for each radionuclide in each sample medium. The reviewer should compare and contrast the pre-operational monitoring program as implemented against the guidance provided in Regulatory Guide 4.14, Revision 1, "Radiological Effluent and Environmental Monitoring at Uranium Mills" (NRC, 1980) and NUREG–5849 (draft), "Manual for Conducting Radiological Surveys in Support of License Termination" (Berger, 1992) or NUREG–1575, Revision 1, "Multi-Agency Radiation Survey and Site Investigation Manual (MARSSIM).

For license renewals and amendment applications, Appendix A to this standard review plan provides guidance for examining facility operations and the approach that should be used in evaluating amendments and renewal applications.

2.9.3   Acceptance Criteria

The characterization of the site background radiological characteristics is acceptable if it meets the following criteria:

(1)   Monitoring programs to establish background radiological characteristics, including sampling frequency, sampling methods, and sampling location and density are established in accordance with pre-operational monitoring guidance provided in Regulatory Guide 4.14, Revision 1, Section 1.1 (NRC, 1980). Air monitoring stations are located in a manner consistent with the principal wind directions reviewed in Section 2.5 of the standard review plan.

(2)   Soil sampling is conducted at both a 5-cm [2-inch] depth as described in Regulatory Guide 4.14, Section 1.1.4 (NRC, 1980) and 15 cm [6 in] for background decommissioning data.

2.9.4   Evaluation Findings

If the staff review, as described in this section, results in the acceptance of the description of the site background radiological characteristics, the following conclusions may be presented in the technical evaluation report and in the environmental assessment.

NRC has completed its review of the characterization information concerned with the background radiological characteristics at the _____ *in situ* leach facility. This review included an evaluation using the review procedures in standard review plan Section 2.9.2 and acceptance criteria outlined in standard review plan Section 2.9.3.

Table 2.9.3-1. Standard Format for Water Quality Data Submittal to the NRC for Uranium Recovery Facilities
1.   Water quality sampling techniques and analysis should be in accordance with U.S. Environmental Protection Agency (EPA) (1974)

2.   All water quality data submitted to NRC should
   a.   Be submitted in tabular form with the appropriate standards (i.e., EPA national interim primary drinking water regulations, livestock standards, baseline or excursion levels, or 10 CFR Part 20, Maximum Permissible Concentrations)[1] listed in the same table, for ease of data comparison. Methods of sampling and preserving and the laboratory utilized should be indicated in the table. The sampled depths, formation(s) sampled, water-level elevations and data measured, and distances from the tailings pond [2] or well field for each monitor should be noted in the table.
   b.   Be submitted graphically to illustrate water quality and water-level elevation changes with time with applicable governing standards, EPA national interim primary drinking water standards and livestock standards, baseline or excursion levels, or maximum permissible concentrations[3] (whatever is appropriate), for the particular constituent on the graph.
   c.   Include a short summary of the data interpretation, noting any anomalies, with an explanation.
   d.   Water quality data reports should include a map that shows all water quality sampling points.

Reference: EPA. "Manual for Chemical Analysis of Water and Wastes". EPA–625–/6–74–003a. Cincinnati, Ohio: EPA, Office of Research and Development Publications. 1974. [1] 10 CFR Part 20 liquid effluent control limits are specified in Table 2 of Appendix B and are not termed Maximum Permissible Concentrations. This table is a direct extraction from the EPA reference. [2] Tailings ponds do not exist at *in situ* leach facilities. This table is a direct extraction from the EPA reference. [3] 10 CFR Part 20 liquid effluent control limits are specified in Table 2 of Appendix B and are not termed Maximum Permissible Concentrations. This table is a direct extraction from the EPA reference.

The licensee has acceptably established the background radiological characteristics by providing (i) monitoring programs to determine background radiologic characteristics that include radionuclides monitored, sampling frequency, and methods, location, and density; (ii) air quality stations located consistent with the prevailing wind directions; (iii) time periods for

Site Characterization

preoperational monitoring that allow for 12 consecutive months of sampling; and (iv) radiologic analyses of soil samples at 5-cm [2-in.] and 15-cm [6-in.] depths.

Based on the information provided in the application, and the detailed review conducted of the characterization of the background radiological characteristics at the _____ *in situ* leach facility, the staff concludes that the information is acceptable to allow evaluation of the radiological background of the site and is in compliance with 10 CFR 51.45, which requires a description of the affected environment containing sufficient data to aid the Commission in its conduct of an independent analysis.

2.9.5 References

Berger, J.D. NUREG/CR–5849, "Manual for Conducting Radiological Surveys in Support of License Termination." Washington, DC: NRC. 1992.

NRC. Regulatory Guide 4.14, "Radiological Effluent and Environmental Monitoring at Uranium Mills." Revision 1. Washington, DC: NRC, Office of Standards Development. 1980.

———. "Multi-Agency Radiation Survey and Site Investigation Manual (MARSSIM)." Revision 1. Washington, DC: NRC. 2000.

EPA. "Manual for Chemical Analysis of Water and Wastes." EPA–625–/6–74–003a. Cincinnati, Ohio: EPA, Office of Research and Development Publications. 1974.

2.10   Other Environmental Features

2.10.1 Areas of Review

This review should include environmental site characterization information that does not clearly fall into any of the other subsections in Section 2 of the standard review plan. These will typically be site-specific, and may be used by the applicant to mitigate unfavorable conditions, or to provide additional information in support of the description of the proposed facility. Information that the applicant believes is important to establish the value of the site and site environs to important segments of the population is appropriately included in this subsection.

2.10.2 Review Procedures

The staff should consider environmental information provided in this section as auxiliary information to support an application for a given facility. The information should be considered in a site-specific context and should be consistent with the information provided in other sections of the application. Depending on the site-specific situation, there may be no information in this section of the application.

For license renewals and amendment applications, Appendix A to this standard review plan provides guidance for examining facility operations and the approach that should be used in evaluating amendments and renewal applications.

2.10.3 Acceptance Criteria

The characterization of other site environmental features is acceptable if it meets the following criteria:

(1)   It is consistent with information provided in previous subsections.

(2)   Information is provided in a manner consistent with good scientific practice, is supported by objective data to the extent possible, and is relevant to the site under consideration.

(3)   Information supports a determination that the *in situ* leach facility can be operated in a manner that will protect public health and safety and the environment.

2.10.4 Evaluation Findings

If the staff review as described in this section results in the acceptance of the description of other environmental features at the site, the following conclusions may be presented in the technical evaluation report and in the environmental assessment.

NRC has completed its review of the characterization information for other environmental features at the _____*in situ* leach facility. This review included an evaluation using the review procedures in standard review plan Section 2.11.2 and acceptance criteria outlined in standard review plan Section 2.11.3.

The licensee has acceptably described any other important environmental features by providing information that is (i) consistent with other aspects of the site description, (ii) supported by objective data, (iii) relevant to the site under consideration, and (iv) supportive of a determination that the *in situ* leach facility can be operated while protecting public health and safety.

Based on the information provided in the application, and the detailed review conducted of the characterization of the other environmental features at the _____ *in situ* leach facility, the staff concludes that the information is acceptable to allow evaluation of the other environmental features, supports associated conceptual and numerical models, and is in compliance with 10 CFR 51.45, which requires a description of the affected environment containing sufficient data to aid the Commission in its conduct of an independent analysis; and 10 CFR Part 40, Appendix A, Criterion 6(7), which provides requirements for control of non-radiological hazards.

2.10.5 References

None.

3.0 DESCRIPTION OF PROPOSED FACILITY

3.1 *In Situ* Leaching Process and Equipment

3.1.1 Areas of Review

The staff should review the *in situ* leaching process as described in the application. This review should include, but not be limited to

(1)   A description of the mineralized zone(s) and the feasibility of processing the defined well field areas

(2)   Well construction techniques and integrity testing procedures to ensure well installations will not result in hydraulic communication between production zones and adjacent non-mineralized aquifers

(3)   A process description including injection/production rates and pressures; plant material balances and flow rates; lixiviant makeup; recovery efficiency; and gaseous, liquid, and solid wastes and effluents that will be generated

(4)   Proposed operating plans and schedules that include timetables and sequences for well field operation, surface reclamation, and ground-water restoration

(5)   Review of techniques for ensuring that a proliferation of small waste disposal sites is avoided.

The review should also include maps showing the facilities layout, descriptions of the process and/or circuit, water and material balances, and the chemical recycling system.

3.1.2 Review Procedures

The staff should determine whether the description of the *in situ* leaching process provided in the application is sufficient to permit evaluation of the operations and processes involved in conformance with the acceptance criteria contained in Section 3.1.3. Staff should ensure the following are included in this section: a map or maps showing the proposed sequence and schedules for uranium extraction and ground-water quality restoration operations, a flow diagram of the process or circuit, a material balance diagram, a description of any chemical recycle systems, a water balance diagram for the entire system, and a map or maps showing the proposed sequence and schedules for land reclamation of the well field areas.

If wells are not properly completed, lixiviant can flow through casing breaks and into overlying aquifers. Casing breaks can occur if the well is damaged during well construction activities. Casing breaks can also occur if water injection pressures exceed the strength of the well materials. Well completion techniques should be reviewed in sufficient detail to give the reviewer a clear understanding of how recovery, injection, and monitor wells are drilled; how their location and spacing are selected; and what materials and methods are used in construction, casing installation, and abandonment. The reviewer should pay particular attention to the techniques employed to prevent hydraulic communication between overlying or

Description of Proposed Facility

underlying aquifers through well boreholes and ensure that secondary ground-water protection standards are not violated (10 CFR Part 40, Appendix A, Criteria 5B, 5C, and 13). Additionally, the applicant should describe methods for well abandonment. The reviewer should ensure that the well casing material used is appropriate for the depths to which the wells are drilled. The reviewer should examine a description of the procedures used to test well integrity. The wells should be retested with sufficient frequency to ensure the integrity of the well construction. The reviewer should examine in detail the justification provided by the licensee for the recommended time interval between successive well integrity tests. The reviewer may refer to a well handbook (e.g., Driscoll, 1989) to verify the appropriateness and expected performance of well installation, testing, and abandonment methods.

To ensure that hydraulic communication between overlying or underlying aquifers through well boreholes is promptly detectable, the reviewer should pay particular attention to the design and installation of vertical and horizontal excursion monitoring wells. Additional review procedures for excursion monitoring systems are provided in Section 5.7.8.2 of this standard review plan.

The reviewer should also pay particular attention to the methods used for effective detection of leaks in surface and near-surface pipes carrying the lixiviant solutions to individual wells within a well field or between the well fields and the processing facilities. Spills of pregnant lixiviant in particular can constitute a significant hazard to health and the environment if allowed to pond and dry on the ground surface, to run off into surface-water bodies, or to infiltrate and transport to ground-water.

The reviewer should determine that any lined impoundment to contain wastes is acceptably designed, constructed, and installed. Materials used to construct the liner should be reviewed to determine that they have acceptable chemical properties and sufficient strength for the design application. The reviewer should determine that the liner will not be overtopped. The reviewer should determine that a proper quality control program is in place. The review should be based on the concept that the site will be in compliance with 10 CFR Part 40, Appendix A, Criterion 2, which precludes long-term disposal of byproduct material onsite and ensures that the proliferation of small waste disposal sites is avoided. The reviewer shall examine the terms of the approved waste disposal agreement.

For surface impoundments containing 11e.(2) byproduct material, the reviewer should ensure that the applicable requirements of 10 CFR Part 40, Appendix A, Criterion 5(A) have been met. If the waste water retention impoundments are located below grade, the reviewer should determine that the surface impoundments have an acceptable liner and leak detection system in place to ensure protection of ground water. The location of a surface impoundment below grade will eliminate the likelihood of embankment failure that could result in any release of waste water. Should the applicant propose to construct a surface impoundment to handle waste water, the reviewer should determine that the design of associated dikes is such that they will not experience massive failure. The design of such dikes to resist erosion and protect against possible flooding events is evaluated in Section 2.7 of this standard review plan. In this section, the reviewer should evaluate the stability of any dikes with respect to seismic events.

In addition, the reviewer should evaluate any proposed surface impoundment to determine if it meets the definition of a dam as given in Regulatory Guide 3.11 (NRC, 1977). If this is the case, the surface impoundment should be included in the NRC Dam Safety Program, and be subject to Section 215, National Dam Safety Program of the Water Resources Development Act of 1996. If the reviewer finds that the impoundment meets the definition of a dam, an evaluation of the dam ranking (low or high hazard) should be made. If the dam is considered a high hazard, an Emergency Action Plan is needed consistent with Federal Emergency Management Agency requirements. For low-hazard dams, no Emergency Action Plan is required. For either ranking of dam, the reviewer should also determine that the licensee has an acceptable inspection program in place to ensure routine checks, and that performance is properly maintained (see Section 5.3 of this standard review plan).

In conducting these evaluations, the reviewer shall consider the technical evaluations conducted by a state or another federal agency with authorities overlapping those of the NRC. Ground-water compliance and protection reviews are the primary technical areas impacted by overlapping authorities. The desired outcome is to identify any areas where duplicative NRC reviews may be reduced or eliminated. The NRC staff must make the necessary evaluations of compliance with applicable regulations for licensing the facility. However, the reviewer may, as appropriate, rely on the applicant's responses to inquiries made by a state or another federal agency to support the NRC evaluation of compliance. The reviewer should make every effort to coordinate the NRC technical review with the state or other federal agency with overlapping authority to avoid unnecessary duplication of effort.

For license renewals and amendment applications, Appendix A to this standard review plan provides guidance for examining historical aspects of facility operations and the approach that should be used in evaluating amendments and renewal applications.

3.1.3 Acceptance Criteria

The *in situ* leaching process and equipment are acceptable if they meet the following criteria:

(1)   The description is sufficiently detailed to identify the mineralized zone(s), their areal distribution, and their approximate thickness.

   If more than one mineralized zone is to be leached, each zone should be defined separately. The estimated U_3O_8 grade should be specified.

(2)   Well design, testing, and inspection reflect accepted NRC practice for *in situ* leach operations.

   (a)   Well Design and Construction—Injection and recovery wells should be constructed from materials that are inert to lixiviants and are strong enough to withstand injection pressures. Polyvinyl Chloride, fiberglass, or acrylonitrile butadiene styrene plastic casings are generally used in wells less than 300-m [1,000-ft] deep. Wells deeper than 300-m [1,000-ft], or those subjected to high-pressure cementing techniques, are subject to collapse. With appropriate design and installation techniques, however, Polyvinyl Chloride can be used for wells greater than 300 m [1,000 ft]. In these instances, steel or fiberglass casing is generally necessary. In all wells (including monitor wells), the annular space between the side of the borehole and the casing should be backfilled with a sealant from the bottom of the casing to the surface in one continuous operation. Proper backfilling isolates the screened formation against vertical migration of water from the surface or from other formations, and also provides support for the casing. Cement or cement-bentonite grout is generally acceptable as a sealant.

Procedures in American Society for Testing and Materials D 5092 provide acceptable methods for design and construction of monitoring wells (American Society for Testing and Materials, 1995). Material normally used for monitor well casing is either metal or plastic. The possibility that chemical reactions may take place between the casing and the mineral constituents in the water affects the choice of casing material used for monitor wells. For example, iron oxide in steel-cased wells will adsorb trace and heavy metals dissolved in the ground water. Therefore, a baseline water sampling program should be used to determine concentrations of trace metals. The applicant should use casing that is inert to these metals, such as Polyvinyl Chloride or fiberglass. When any well is completed, it should be developed until production of essentially sediment-free water is assured for the life of the well. One acceptable development method is to use a swab in the well to create a vacuum on the upstroke and positive pressure on the downstroke. Air lifting is also an acceptable method for well development. Other state- or EPA-approved well development methods may also be used.

(b) Well Integrity Testing—Injection and recovery wells should be tested for mechanical integrity. The following are examples of well integrity testing procedures that have been considered acceptable in previous applications. To inspect for casing leaks after a well has been completed and opened to the aquifer, a packer is set above the well screen, and each well casing is filled with water. At the surface, the well is pressurized with either air or water to 125 percent of the maximum operating pressure. The well pressure is then monitored for a period of 10 minutes to 20 minutes, with a pressure drop of no more than 10 percent, to ensure significant pressure drops do not occur through borehole leaks. Operating pressure varies with the depth of the well and should be less than formation fracture pressure. Well integrity tests should be performed on each injection and production well before the wells are utilized and on wells that have been serviced with equipment or procedures that could damage the well casing. Additionally, each well should be retested with sufficient frequency (once each 5 years or less) to ensure the integrity of the well construction if it is in use. Sole reliance on single-point resistance geophysical tools is not acceptable for determining the mechanical integrity at a well.

(3) The number, location, and screened intervals of excursion monitoring wells are described in sufficient detail, follow industry standard practice, and are adequate to ensure prompt detection of horizontal and vertical excursions, taking into account site specific parameters such as local geology and hydrology. Acceptance criteria for methods and calculations used to determine the placement of horizontal and vertical

excursion monitoring wells are presented in Section 5.7.8.3 of this standard review plan.

(4)   Methods for timely detection and cleanup of leaks from surface and near-surface pipes within the well fields and between the well field and processing facilities are clearly described and included in the design.

(5)   The description of the *in situ* leaching process includes the following information and demonstrations:

(a)   Projected down-hole injection pressures with the hydrostatic pressure of the fluid column should be demonstrated to be maintained below casing (casing and cement) failure pressures and formation fracture pressures, to avoid hydrofracturing the aquifer and promoting leakage into the overlying units. Piping burst strength should be considered in deep well fields {greater than about 305 m [1,000 ft]}.

(b)   Overall production rates should be higher than injection rates.

(c)   Proposed plant material balances and flow rates should be acceptably described.

(d)   Lixiviant makeup should be such that impact on the ground-water quality and the prospects for long-term ground-water restoration will be maintained at levels that ensure acceptable restoration goals can be achieved in a timely manner. Oxidants such as gaseous oxygen and hydrogen peroxide, and carbonates such as sodium bicarbonate or carbon dioxide gas have been demonstrated in a number of *in situ* leach facilities to be suitable lixiviants.

(e)   The description should identify gaseous, liquid, and solid wastes and effluents that will be generated. Effluent monitoring and control measures are discussed in Section 4.0 of this standard review plan.

(f)   An analysis of the effects that *in situ* leach operations are likely to have on surrounding water users has been provided. An acceptable impact analysis should be based on results of numerical or analytical modeling calculations that are used to estimate ground-water travel times from the proposed extraction areas to the nearby points of ground-water or surface-water usage, estimate the amount of process bleed necessary to prevent migration of lixiviant from the well field, and describe the applicant's mitigative measures to recover lixiviant excursions. If the applicant chooses to use nominal parameter estimates, parameter uncertainties should be considered to ensure that the selected values represent expected conditions. An acceptable impact analysis should describe the following:

(i)   The ability to control the migration of lixiviant from the production zones to the surrounding environs

3-5

(ii)   Ground-water and surface-water pathways that might transport extraction solutions offsite in the event of an uncontrolled excursion, surface piping leak, or incomplete restoration

(iii)   The impact of *in situ* leach operations on ground-water flow patterns and aquifer levels

(iv)   The expected post-extraction impact on geochemical properties and water quality

(6)   Proposed operating plans and schedules include timetables for well field operation, surface reclamation, and ground-water restoration. Water balance calculations should be provided that demonstrate that the liquid waste disposal facilities (surface impoundments, land application, deep well injection) are adequate to process the proposed production and restoration efforts at any time.

(7)   The staff should verify the applicant analyses or perform independent review analyses of floods and flood velocities. If the design assumptions and calculations are reasonable, accurate, and compare favorably with independent staff estimates, the designs are acceptable.

(8)   The staff should evaluate the design of diversion channels in several critical areas using the criteria and guidance presented in NUREG–1623 (NRC, 1998). For the main channel area, the staff should verify that appropriate models and input parameters have been used to design the erosion protection. The staff should assure that flow rates, flow depths, and shear stresses have been correctly computed. The diversion channels should be sized and protected to pass a probable maximum flood with minimal, if any, damage to the diversion channel. No release of contained materials should occur during a probable maximum flood. The staff should determine that the depth of burial of any disposed of material is sufficient to preclude bottom scouring, if an existing or constructed channel is located in or near a pit or impoundment. Where practical, the use of diversion channels at new facilities should be avoided to lessen costs of reclamation and future maintenance.

(9)   The staff should review the plans, specifications, inspection programs, and quality assurance/quality control programs to assure that acceptable measures are being taken to construct the facility according to accepted engineering practices. The staff will compare the information provided with typical programs used in the construction industry.

(10)   Results from research and development or other production operations are used to support the description of the *in situ* leaching process, where appropriate.

(11)   The applicant has an approved waste disposal agreement for 11e.(2) byproduct material disposal at an NRC or NRC Agreement State licensed disposal facility. This agreement is maintained onsite. The applicant has committed to notify NRC in writing within 7 days if this agreement expires or is terminated and to submit a new agreement for NRC

approval within 90 days of the expiration or termination (failure to comply with this license condition will result in a prohibition from further lixivient injection).

3.1.4   Evaluation Findings

If the staff review as described in this section results in the acceptance of the *in situ* leaching process and equipment, the following conclusions may be presented in the technical evaluation report.

NRC has completed its review of the *in situ* leaching process and equipment proposed for use at the _____ *in situ* leach facility. This review included an evaluation using the review procedures in standard review plan Section 3.1.2 and the acceptance criteria in standard review plan Section 3.1.3.

The applicant has acceptably described the mineralized zone(s) demonstrated protection against vertical migration of water, proposed tests for well integrity, and demonstrated that the *in situ* leaching process will meet the following criteria: (i) down hole injection pressures are less than formation fracture pressures; (ii) overall production rates are higher than injection rates; (iii) plant material balances and flow rates are appropriate; (iv) lixiviant makeup is such that restoration goals can be achieved in a timely manner; (v) recovery efficiency is assessed through mass balance calculations; and (vi) reasonable estimates of gaseous, liquid, and solid wastes and effluents are provided (used in evaluation of effluent monitoring and control measures in standard review plan Section 4.0). The applicant has used the results from research and development or other production operations to support the evaluation of the *in situ* leaching process. The applicant has provided acceptable operating plans, schedules, and timetables for well field operation, surface reclamation, and ground-water restoration.

Based on the information provided in the application and the detailed review conducted of the *in situ* leaching process and equipment for the _____ *in situ* leach facility, the staff concludes that the proposed *in situ* leaching process and equipment are acceptable and are in compliance with 10 CFR 40.32(c), which requires the applicant's proposed equipment, facilities, and procedures to be adequate to protect health and minimize danger to life or property; 10 CFR 40.32(d), which requires that the issuance of the license will not be inimical to the common defense and security or to the health and safety of the public; 10 CFR 40.41(c), which requires the applicant to confine source or byproduct material to the location and purposes authorized in the license; and 10 CFR Part 40, Appendix A, Criteria 2 for non-proliferation of small disposal sites; 5(A) for ground-water protection; 5B for secondary ground-water protection; 5C for maximum values for ground-water protection; and 13 for hazardous constituents. The related reviews of the 10 CFR Part 20 radiological aspects of the *in situ* leaching process and equipment in accordance with standard review plan Sections 4.0, "Effluent Control Systems;" 5.0, "Operations;" and 7.0, "Environmental Effects;" are addressed elsewhere in this technical evaluation report.

Description of Proposed Facility

3.1.5 References

American Society for Testing and Materials. "Standard Practice for Design and Installaiton of Ground Water Monitoring Wells in Aquifers." Designation D5092-90. Philadelphia, Pennsylvania: American Society for Testing and Materials. 1995.

Driscoll, F.G. "Groundwater and Wells." St. Paul, Minnesota: Johnson Filtration Systems, Inc. 1989.

NRC. "Recommendation on Ways to Improve the Efficiency of NRC Regulation at *In Situ* Leach Uranium Recovery Facilities." SECY–99–0013. Washington, DC: NRC. 2000.

———. NUREG–1623, "Design of Erosion Protection for Long-Term Stabilization." Washington, DC: NRC. 1998.

———. Regulatory Guide 3.11, "Design, Construction, and Inspection of Embankment Retention Systems for Uranium Mills." Washington, DC: NRC, Office of Standards Development. 1977.

3.2 Recovery Plant, Satellite Processing Facilities, Well Fields, and Chemical Storage Facilities—Equipment Used and Materials Processed

3.2.1 Areas of Review

The staff should review the physical descriptions and reported operating characteristics for the major equipment items of the processing cycle. The staff should also review descriptions of the proposed process information and controls, as well as radiation sampling and monitoring equipment. Controls mean the apparatus or mechanisms that could affect the chemical, physical, metallurgical, or nuclear processes of the facility in such a manner as to influence radiation health and safety. The staff should review a diagram that indicates the plant layout and locations where dusts, fumes, or gases would be generated; locations of all ventilation, filtration, confinement, and dust collection systems; and radiation safety and radiation monitoring devices.

In addition, staff should review the list and specifications related to all radioactive and hazardous materials used in the recovery plant, satellite processing facilities, well fields, and chemical storage facilities. These should be reviewed for the hazards associated with the quantities, locations, operating flow rates, temperatures, and pressures associated with these materials.

While safety concerns with the use of all hazardous materials are important and need to be addressed, direct NRC regulatory authority is limited to situations where hazardous materials have a potential affect on radiological safety. Chemicals of concern typically used in the uranium *in situ* leach facilities are identified in NUREG/CR–6733 (NRC, 2001). Therefore, staff should review the list of applicable federal, state, and local regulations that the licensee intends to use, to ensure that all hazardous chemicals that have the potential to impact radiological safety, are safely handled. Staff should also review the safety features used in the facility process design for eliminating or mitigating the hazards presented by these materials.

3.2.2 Review Procedures

The staff should determine whether the physical descriptions and reported operating characteristics for the major equipment items of the processing cycle, the proposed controls, and safety/radiation instrumentation are sufficient to evaluate the performance of the proposed uranium *in situ* leach facility. Staff should ensure that the application identifies all areas where releases of radioactive and hazardous materials (such as radon gas and uranium dust) can occur and that locations of control equipment (e.g., ventilation and exhaust systems) and instrumentation are provided.

Staff should determine whether the hazards associated with the storage and processing of the radioactive materials and those hazardous materials with the potential to impact radiological safety, have been sufficiently addressed in the process design for the recovery plant, satellite processing facilities, well fields, and chemical storage facilities.

For license renewals and amendment applications, Appendix A to this standard review plan provides guidance for examining facility operations and the approach that should be used in evaluating amendments and renewal applications.

3.2.3 Acceptance Criteria

The description of the equipment used and materials processed in the recovery plant, satellite processing facilities, well fields, and chemical storage facilities is acceptable if it meets the following criteria:

(1)   The application provides diagrams showing the proposed (or existing) plant/facilities layout in adequate detail.

(2)   Areas where dusts, fumes, or gases would be generated are clearly identified, along with a description of the source of the emissions.

(3)   All ventilation, filtration, confinement, dust collection, and radiation monitoring equipment are described as to size, type, and location.

(4)   Availability requirements for safety equipment are adequately stated, and measures for ensuring availability and reliability are clearly identified.

(5)   Specifications, quantities, locations, and operating conditions such as flow rates, temperatures, and pressures of radioactive materials and those hazardous materials with the potential to impact radiological safety, are clearly identified together with the hazards associated with these materials.

(6)   A list of applicable federal, state and local regulations that the licensee intends to use to ensure that process chemicals having the potential to impact radiological safety are

Description of Proposed Facility

safely handled, is provided.

(7)   Controls used for eliminating or mitigating the hazards presented by the radioactive materials and those hazardous materials with the potential to impact radiological safety, are adequately described.

Further discussion on Criteria 4–7 may be found in NUREG/CR–6733 (NRC, 2001).

3.2.4   Evaluation Findings

If the staff review as described in this section results in the acceptance of the equipment used and materials processed in the *in situ* leach facility, the following conclusions may be presented in the technical evaluation report.

NRC has completed its review of the equipment proposed for use and materials to be processed in the recovery plant, satellite processing facilities, well fields, and chemical storage facilities at the _____ *in situ* leach facility. This review included an evaluation using the review procedures in standard review plan Section 3.2.2 and the acceptance criteria outlined in standard review plan Section 3.2.3.

Based on the information provided in the application and the detailed review conducted of the equipment to be used and materials to be processed in the recovery plant, satellite processing facilities, well fields and chemical storage facilities for the _____ *in situ* leach facility, the staff concludes that the proposed equipment to be used and materials to be processed in the recovery plant, satellite processing facilities, well fields, and chemical storage facilities are acceptable and are in compliance with 10 CFR 40.32(c), which requires that applicant proposed equipment, facilities, and procedures be adequate to protect health and minimize danger to life or property; 10 CFR 40.32(d), which requires that the issuance of the license will not be inimical to the common defense and security or to the health and safety of the public; and 10 CFR 40.41(c), which requires the applicant to confine source or byproduct material to the locations and purposes authorized in the license. The related reviews of the 10 CFR Part 20 radiological aspects of the recovery plant equipment in accordance with standard review plan Sections 4.0, "Effluent Control Systems;" 5.0, "Operations;" and 7.0, "Environmental Effects" are addressed elsewhere in this technical evaluation report.

3.2.5   Reference

NRC. NUREG/CR–6733, "A Baseline Risk-Informed, Performance-Based Approach for *In Situ* Leach Uranium Extraction Licensees." Washington, DC: NRC. 2001.

3.3   Instrumentation and Control

3.3.1   Areas of Review

The staff should review descriptions of the proposed process instrumentation and controls and radiation safety sampling and monitoring instrumentation, including their minimum specifications and operating characteristics. This review should include well field process control equipment for monitoring injection pressures, injection rates, and production rates. It should also include safety related process monitoring and control equipment used in the recovery plant, satellite processing facilities, well fields, chemical storage facilities, and surface impoundments.

3.3.2 Review Procedures

The staff should review the descriptions of the proposed instrumentation and control systems provided in the application to determine whether they are sufficient to evaluate the interrelationship between the proposed instrumentation systems and the operations or processes to be controlled or monitored. The staff should also determine whether the proposed instrumentation systems are sufficient to control and monitor operations and processes identified in the description of the proposed facility. Particular attention should be focused on whether proposed monitoring and control instrumentation is adequate to quickly identify and remedy *in situ* leaching and processing problems that can increase exposures to radiological and chemical hazards. Areas of concern include monitoring and ventilation systems designed to detect and control elevated releases of yellowcake dust from drying and storage operations and radon gas buildup in buildings. Areas of concern also include instrumentation used to record, monitor and control key operating parameters of the yellowcake dryers and their associated stack emission scrubbing systems. Instrumentation to detect and control liquid releases from well field and processing pipe failures, surface impoundment leaks, and chemical tank valve failures should also be evaluated in the staff review.

For license renewals and amendment applications, Appendix A to this standard review plan provides guidance for examining facility operations and the approach that should be used in evaluating amendments and renewal applications.

3.3.3 Acceptance Criteria

The facility instrumentation is acceptable if it meets the following criteria:

(1)   Instrumentation has been described for the various components of the processing facility, including well fields, well field houses, trunk lines, the production circuit, surface impoundments, and deep injection disposal wells.

(2)   Instrumentation is designed to allow the plant operator to continuously monitor and control a variety of systems and parameters, including total flow into the plant, total waste flow leaving the plant, tank levels, and the yellowcake dryer. Instrumentation includes alarms and interlocks in the event of a failure.

(3)   Control components of the systems are equipped with backup systems that activate in the event of a failure of the operating system or a common cause failure such as power failure.

(4)   Well field operating pressures are kept below casing and formation rupture pressures to prevent vertical excursions. Well field operation pressures are routinely monitored

Description of Proposed Facility

        either at the well head or on the entire system, and are measured and recorded daily.

(5)   Manufacturer's recommendations for maintenance and operation of yellowcake dryers, and checking and logging requirements contained in 10 CFR Part 40, Appendix A. Criterion 8 are followed.

3.3.4 Evaluation Findings

If the staff review as described in this section results in the acceptance of the facility instrumentation and control systems, the following conclusions may be presented in the technical evaluation report.

NRC has completed its review of the instrumentation and control proposed for use at the _____ *in situ* leach facility. This review included an evaluation using the review procedures in standard review plan Section 3.3.2 and the acceptance criteria outlined in standard review plan Section 3.3.3.

The instrumentation and control systems have been acceptably described for components including the well fields, well field houses, trunk lines, production circuit, surface impoundments, and deep injection disposal wells. The instrumentation allows for continuous monitoring and control of systems, including total inflow to the plant, total waste flow exiting the plant, tank levels, and the yellowcake dryer. Appropriate alarms and interlocks are part of the instrumentation systems. Each control system is equipped with an acceptable backup system that automatically activates in the event of a failure of the operating system or a common cause failure such as a power failure.

Based on the information provided in the application and the detailed review conducted of the instrumentation and control for the _____ *in situ* leach facility, the staff concludes that the proposed instrumentation is acceptable and is in compliance with 10 CFR 40.32(c), which requires applicant proposed equipment, facilities, and procedures to be adequate to protect health and minimize danger to life or property; 10 CFR 40.32(d), which requires that the issuance of the license will not be inimical to the common defense and security or to the health and safety of the public; and 10 CFR 40.41(c), which requires the applicant to confine source or byproduct material to the locations and purposes authorized in the license. The related reviews of the 10 CFR Part 20 radiological aspects of the solution mining process and equipment, in accordance with standard review plan Sections 4.0, "Effluent Control Systems;" 5.0, "Operations;" and 7.0, "Environmental Effects" are addressed elsewhere in this technical evaluation report.

3.3.5 References

None.

4.0 EFFLUENT CONTROL SYSTEMS

4.1 Gaseous And Airborne Particulates

4.1.1 Areas of Review

The staff should review the proposed ventilation, filtration, and confinement systems that are to be used to control the release of radioactive materials to the atmosphere. The staff should also review analyses of equipment as designed and operated to prevent radiation exposures and to limit exposures and releases to as low as is reasonably achievable. A review should also be conducted of a physical description of discharge stacks, types and estimated composition and flow rates of atmospheric effluents, and proposed methods for controlling such releases.

4.1.2 Review Procedures

The staff should review facilities, designs, and operational modes to determine whether the proposed ventilation, filtration, and confinement systems and equipment described in the application are sufficient to control the release of radioactive materials to the atmosphere to meet acceptance criteria identified in Section 4.1.3.

4.1.3 Acceptance Criteria

The gaseous and airborne particulate effluent control systems are acceptable if they meet the following criteria:

(1)   Monitoring and control systems for the facility are located to optimize their intended function. Monitors used to assess worker exposures are placed in locations of maximum anticipated concentration based upon determination of airflow patterns.

(2)   Monitoring and control systems for the facility are appropriate for the types of effluents generated. The intended purposes of measurement devices are clearly stated and criteria for monitoring are provided. The acceptance criteria from Section 5.7.7.3 of this standard review plan should be met.

(3)   The application provides a demonstration that adequate ventilation systems are planned for process buildings to avoid radon gas buildup. Ventilation systems should be consistent with the requirements of Regulatory Guide 8.31, "Information Relevant to Ensuring that Occupational Radiation Exposures at Uranium Mills Will Be as Low as Is Reasonably Achievable," Section 3.3 (NRC, 2002).

The review emphasis should be on radon gas mobilization from (i) recovery solutions entering the plant, (ii) the extraction process (where tanks are vented), and (iii) uranium particulate emissions resulting from drying and packaging operations and spills. For facilities using an open air design for processing (i.e., processing equipment is not enclosed by a building), ventilation will be less of a safety concern. Aspects of design that can significantly limit airborne releases include closed production systems (i.e., no venting) and the use of vacuum dryers that eliminate airborne uranium particulate releases from drying operations.

Effluent Control Systems

(4)   The application demonstrates that the effluent control systems will limit exposures under both normal and accident conditions. The application also provides information on the health and safety impacts of system failures and identifies contingencies for such occurrences.

(5)   The application demonstrates that the operations will be conducted so that all airborne effluent releases are as low as is reasonably achievable.

4.1.4   Evaluation Findings

If the staff review as described in this section results in the acceptance of the effluent control systems for gaseous and airborne particulates, the following conclusions may be presented in the technical evaluation report and environmental assessment.

NRC has completed its review of the effluent control systems for gaseous and airborne particulates proposed for use at the _____ *in situ* leach facility. This review included an evaluation using the review procedures in standard review plan Section 4.1.2 and the acceptance criteria outlined in standard review plan Section 4.1.3.

The applicant has acceptably described the discharge stacks and the types, estimated composition, and flow rates of effluents released to the atmosphere. The applicant has designated monitoring and control systems (e.g., ventilation, filtration, and confinement) for the types of effluents generated. Also, the applicant has specified acceptable monitoring criteria and has located the facility monitoring and control systems for the required functions to optimally assess worker exposure in locations of likely maximum concentrations determined by the applicant's analysis of airflow patterns. The applicant has demonstrated that ventilation systems are acceptable to prevent radon gas buildup where (i) recovery solutions enter the plant, (ii) tanks are vented during the extraction process, and (iii) drying and packaging operations occur. By providing information on the health and safety impacts of system failures and identifying contingencies for such occurrences, the applicant has acceptably shown that effluent control systems will limit radiation exposures under both normal and accident conditions. The applicant has committed to occupational radiation doses and doses to the general public that meet dose limits and as low as is reasonably achievable goals.

Based on the information provided in the application and the detailed review conducted of the effluent control systems for gaseous and airborne particulates for the _____ *in situ* leach facility, the staff concludes that the proposed effluent control systems for gaseous and airborne particulates are acceptable and are in compliance with 10 CFR 20.1101, which requires that an acceptable radiation protection program that achieves as low as is reasonably achievable goals is in place and that a constraint on air emissions, excluding Radon-222 and its decay products, will be established to limit doses from these emissions; 10 CFR 20.1201, which defines the allowable occupational dose limits for adults; 10 CFR 20.1301, which defines dose limits allowable for individual members of the public; 10 CFR 20.1302, which requires compliance with dose limits for individual members of the public; 10 CFR Part 40, Appendix A, Criterion 5(G)(1), which requires that the chemical and radioactive characteristics of wastes be defined; and 10 CFR Part 40, Appendix A, Criterion 8, which provides requirements for control of airborne effluent releases. The related reviews of the 10 CFR Part 20 radiological aspects of the effluent control systems for gaseous and airborne radionuclides in accordance with standard review plan Sections 5.0, "Operations;" and 7.0, "Environmental Effects" are addressed elsewhere in this technical evaluation report.

4.1.5 Reference

NRC. Regulatory Guide 8.31, "Information Relevant to Ensuring that Occupational Radiation Exposures at Uranium Mills will be as low as is Reasonably Achievable." Washington, DC: NRC, Office of Standards Development. 2002.

4.2 Liquids and Solids

4.2.1 Areas of Review

The staff should review estimates of quantities and compositions of waste residues expected during construction and operation and the procedures proposed for their management. The staff should also review design specifications for effluent control systems for liquids and solids. Staff should review the design specifications of any retention systems such as surface impoundments. If effluents are to be released into surface waters or injected into disposal wells, the staff should also review the plans to obtain any water quality certifications and discharge permits that may be necessary.

Areas to be reviewed include

(1)   Information related to surface impoundment design, monitoring programs, freeboard requirements, and leak reporting procedures

(2)   Liquid effluent disposal plans

(3)   Contingency plans for dealing with leaks and spills

(4)   Contaminated solid waste generation and disposal plans

(5)   Non-contaminated solid waste generation and disposal plans

4.2.2 Review Procedures

The staff should ensure that facility descriptions include a discussion of design features to contain contamination from spills resulting from normal operations and the likely consequences of any accidents (e.g., valve and tank failures, leaks in impoundment liners). The staff should perform the following assessments:

(1)   Verify that surface impoundments rely on standard engineering design to ensure proper containment performance, including appropriate leak detection systems. The staff should also ensure that appropriate freeboard requirements are established, and that appropriate monitoring programs and reporting procedures are in place.

Effluent Control Systems

(2)   If liquid effluents are to be released into surface waters, applied to land surfaces, or injected into disposal wells, determine whether the applicant has applied for or been issued appropriate water quality certifications and discharge permits (see standard review plan Section 10.0 for review of these documents). If the applicant has not yet applied for or been issued such permits, the reviewer should determine that the applicant has identified the necessary permits, and should ensure that a license condition is required prohibiting mineral extraction until all permits are received.

(3)   Ensure that contingency plans are in place for dealing with spills of process fluids from valve, pipe, or tank failures.

(4)   Ensure that an agreement is in place for disposal of 11.e(2) byproduct material in an NRC licensed disposal facility or a licensed mill tailings facility.

In evaluating surface impoundments, an evaluation of environmental impacts must be made, and a conclusion of the acceptability of those impacts should be documented. The reviewer should also determine if the design of the impoundment meets the applicable requirements of 10 CFR Part 40, Appendix A.

4.2.3 Acceptance Criteria

The liquids and solids effluent control systems are acceptable if they meet the following criteria:

(1)   Common liquid effluents generated from the process bleed, process solutions (e.g., backwash, resin transfer waters), wash-down water, well development water, pumping test water, and restoration waters are properly controlled.

Acceptable control methods include diversion of liquid wastes to surface impoundments, deep well injection, and land application/irrigation. Solid effluents can be considered either as contaminated or as noncontaminated. Contaminated solid effluent that can be decontaminated and released for unrestricted use is discussed in detail in Section 5.7.6 of this standard review plan.

To dispose of liquid waste by on-site land application, the applicant must provide (i) a description of the waste including its physical and chemical properties that are important to risk, (ii) a description of the proposed manner and conditions of waste disposal, (iii) an analysis and evaluation of pertinent information on the affected environment, (iv) information on the nature and location of other facilities likely to be affected, and (v) analyses and procedures to ensure that doses are maintained as low as is reasonably achievable and within the dose limits in 10 CFR 20.1301.

For land application, the applicant must analyze and assess projected (i) concentrations of radioactive contaminants in the soils to show that the concentration of radium and other nuclides in the soil will not exceed the standard in 10 CFR Part 40, Appendix A, Criterion 6(6); (ii) impacts on ground-water and surface-water quality; (iii) impacts on land use, particularly crops and vegetation; and (iv) exposures and health risks that may be associated with radioactive constituents reaching the food chain. All projected doses and risks must conform to the risk levels permitted under 10 CFR Part 20. The applicant should propose periodic soils surveys that include contaminant monitoring to verify that contaminant levels in the soil do not exceed the projected levels. A remediation plan must be in place to be implemented in the event that the projected levels are exceeded.

The applicant must conduct analyses to assess the chemical toxicity of radioactive and nonradioactive constituents to evaluate health risks associated with land application involving irrigation at particular sites. The staff should determine that the specific toxicity evaluations and any necessary permits are sufficient to conform to the applicable regulations such as 10 CFR 20.2007. In the absence of compliance monitoring wells in the uppermost aquifer in the area used for land application, the applicant must demonstrate that contaminants will not be returned to the ground water and cause any exceedance of site-specific ground-water protection standards.

Applicants are required to comply with NRC requirements for decommissioning before facility closure and license termination. (Decommissioning requirements are discussed in Section 6 of this standard review plan.)

(2)   On-site evaporation systems are designed and operated in a manner that prevents migration of waste from the evaporation system to the subsurface.

The following discussion provides guidelines for an acceptable application section dealing with surface impoundments.

The monitoring and inspection program consists of documented daily checks of impoundment freeboard and the leak detection system. Because small amounts of condensation can accumulate in leak detection sumps, samples for chemical analysis are not commonly collected until water levels greater than a specified amount are detected. NRC has found 15 cm [6 in.] to be an acceptable level. When significant water levels are detected, the water in the standpipes must be sampled for indicator parameters to confirm that the water in the detection system is from the impoundment. The applicant should specify and provide the basis for selecting the indicator parameter(s) used to verify leaks.

Corrective actions should commence on leak confirmation and should consist of transferring the solution to another impoundment so that liner repairs can be made. Thus, sufficient freeboard capacity should be maintained in the surface impoundments such that any one impoundment could be transferred to the remaining impoundments in the event of a leak. An additional freeboard requirement is that water levels should be kept far enough below the top of the impoundment to prevent waves from overtopping during high wind conditions.

Actions to be taken in the event that surface impoundment water analyses indicate leakage include (i) notifying NRC by telephone within 48 hours of verification, (ii) analyzing standpipe water quality samples for leak parameters once every 7 days during the leak period and once every 7 days for at least 14 days following repairs, and

(iii) filing a written report with NRC within 30 days of first notifying NRC that a leak exists. (This report includes analytical data and describes the corrective actions and the results of those actions.)

(3)   The design, installation, and operation of surface impoundments at the site used to manage 11e.(2) byproduct material meet relevant guidance provided in Regulatory Guide 3.11, Section 1 (NRC, 1977). The impoundments should have sufficient capacity that the entire contents of one impoundment can be transferred to the other surface impoundments in the event of a leak. (See Section 2.7.3 of this standard review plan for additional discussion of design and evaluation of retention systems and diversion facilities.) Inspections of impoundments will be done consistent with Regulatory Guide 3.11.1, "Operational Inspection and Surveillance of Embankment Retention Systems for Uranium Mill Tailings" (NRC, 1980).

The surface impoundment must have sufficient capacity and must be designed, constructed, maintained, and operated to prevent overtopping resulting from (i) normal or abnormal operations, overfilling, wind and wave actions, rainfall, or run-on; (ii) malfunctions of level controllers, alarms, and other equipment; and (iii) human error. If dikes are used to form the surface impoundment, the dikes must be designed, constructed, and maintained with sufficient structural integrity to prevent massive failure of the dikes. In ensuring structural integrity, the applicant must not assume that the liner system will function without leakage during the active life of the impoundment.

Controls should be established over access to the impoundment, including access during routine maintenance. A procedure should be provided that assures that unnecessary traffic is not directed to the impoundment area.

(4)   The design of surface impoundments used in the management of 11e.(2) byproduct material meets or exceeds the requirements in 10 CFR Part 40, Appendix A, Criterion 5(A).

The design of a clay or synthetic liner and its appurtenant component parts should be presented in the application or related amendment applications for a uranium recovery operation. At a minimum, design details, drawings, and pertinent analyses should be provided. Expected construction methods, testing criteria, and quality assurance programs should be presented. Planned modes of operation, inspection, and maintenance should be discussed in the application. Deviation from these plans should be submitted to and approved by the staff before implementation.

The liner for a surface impoundment used to manage 11e.(2) byproduct material must be designed, constructed, and installed to prevent any migration of wastes out of the impoundment to the subsurface soil, ground-water, or surface-water at any time during the active life of the surface impoundment. The liner may be constructed of materials that allow wastes to migrate into the liner provided that the impoundment decommissioning includes removal or decontamination of all waste residues, contaminated containment system components, contaminated subsoils, and structures and equipment contaminated with waste and leachate.

The liner must be constructed of materials that have appropriate chemical properties and sufficient strength and thickness to prevent failure because of pressure gradients, physical contact with the waste or leachate, climatic conditions, and the stresses of installation and daily operation. The subgrade must be sufficient to prevent failure of the liner because of settlement, compression, or uplift. Liners must be installed to cover all surrounding earth which is likely to be in contact with the wastes or leachate.

Tests should show conclusively that the liner will not deteriorate when subjected to the waste products and expected atmospheric and temperature conditions at the site. Applicant test data and all available manufacturers test data should be submitted with the application. For clay liners, tests, at a minimum, should consist of falling head permeameter tests performed on columns of liner material obtained during and after liner installation. The expected reaction of the impoundment liner to any combination of solutions or atmospheric conditions should be known before the liner is exposed to them. Field seams of synthetic liners should be tested along the entire length of the seam. Representative sampling may be used for factory seams. The testing should use state-of-the-art test methods recommended by the liner manufacturer. Compatibility tests that document the compatibility of the field seam material with the waste products and expected weather conditions should be submitted for staff review and approval. If it is necessary to repair the liner, representatives of the liner manufacturer should be called on to supervise the repairs.

Proper preparation of the subgrade and slopes of an impoundment is very important to the success of the surface impoundment. The strength of the liner is heavily dependent on the stability of the slopes of the subgrade. The subgrade should be treated with a soil sterilant. The subgrade surface for a synthetic liner should be graded to a surface tolerance of less than 2.54 cm [1 in.] across a 30.3 cm [1 ft] straightedge. NRC Regulatory Guide 3.11, Section 2 (NRC, 1977) outlines acceptable methods for slope stability and settlement analyses, and should be used for design. If a surface impoundment with a synthetic liner is located in an area where the water table could rise above the bottom of the liner, under drains may be required. The impoundment will be inspected in accordance with Regulatory Guide 3.11.1 (NRC, 1980).

A quality control program should be established for the following factors: (i) clearing, grubbing, and stripping; (ii) excavation and backfill; (iii) rolling; (iv) compaction and moisture control; (v) finishing; (vi) subgrade sterilization; and (vii) liner subdrainage and gas venting.

To prevent damage to liners, some form of protection should be provided, including (i) soil covers, (ii) venting systems, (iii) diversion ditches, (iv) side slope protection, or (v) game-proof fences. A program for maintenance of the liner features should be developed, and repair techniques should be planned in advance.

A leak detection system should be installed at all sites using natural or synthetic liners. The system should be designed to perform the following functions: (i) detect accidental leaks from the impoundment, (ii) identify the location of the leak so that liner repair can be implemented immediately, and (iii) isolate the leakage and control it.

Effluent Control Systems

Inspections should be made of the liner, liner slopes, and other earthwork features. Any damage or defects that could result in leakage should be immediately reported to the staff. Appropriate repairs should be implemented as soon as possible.

(5)   Plans and procedures are provided for addressing contingencies for all reasonably expected system failures and include:

   (a)   A listing of the likely consequences of any failures in process or well field equipment that could result in a release of material

   (b)   Identification of appropriate plant and corporate personnel who must be notified in the event of specific types of failures

   (c)   Measures for quickly containing and mitigating the impacts of released materials

   (d)   Provisions for issuing radiation work permits for workers to mitigate impacts

   (e)   Specific procedures for complying with notification requirements in the regulations, license, and other permits, as appropriate

   Processing plants should have sump capacity sufficient to contain the volume of the largest tank in the plant that contains hazardous material. Well field flow circuits should be equipped with alarms to notify the operator in the event of loss of pressure or excess pressure anywhere within the production circuit. NRC should be notified of spills in accordance with criteria in Section 5.3.1.3(2) of this standard review plan.

(6)   The application contains a description of the methods to be used for disposing of contaminated solid wastes that are generated during operation of the facility. Decommissioning wastes are addressed separately in Chapter 6 of this review plan.

   Equipment that can be decontaminated and released for unrestricted use is discussed in Section 5.7.6 of this standard review plan. The storage of byproduct material that either cannot or will not be decontaminated and released for unrestricted use will be managed to ensure compliance with occupational dose limits in 10 CFR Part 20, Subpart C. The detailed review of occupational doses will be completed as described in Section 5.7 of this standard review plan. The application should provide an estimate of the amount of contaminated material that will be generated and objective evidence of an agreement for disposal of these materials either in a licensed waste disposal site or at a licensed mill tailings facility.

   The applicant has an approved waste disposal agreement for 11e.(2) byproduct material disposal at an NRC or NRC Agreement State licensed disposal facility. This agreement is maintained onsite. The applicant has committed to notify NRC in writing within 7 days if this agreement expires or is terminated and to submit a new agreement for NRC approval within 90 days of the expiration or termination (failure to comply with this license condition will result in a prohibition from further lixivient injection).

(7) Water quality certification and discharge permits have been obtained, or plans are in place to obtain them (review requirements for the status of these permits are addressed in Section 10.0 of the standard review plan). If such permits are not yet applied for or issued, the reviewer should determine that the applicant has identified the necessary permits and should ensure that a license condition is required prohibiting lixiviant injection until all permits are received. Table 4.2.3-1 provides a list of non-NRC permits that may be required to support liquid effluent disposal at *in situ* leach facilities.

(8) Acceptable methods for effluent disposal by release to surface water, evaporation from surface impoundments, land application, and deep well injection are consistent with NRC guidance.

(9) Alternatives to liquid management activities have been considered and none is found to be obviously superior to the selected option. In addition, environmental impacts from all liquid waste management activities have been found to be acceptable.

4.2.4 Evaluation Findings

If the staff review as described in this section results in the acceptance of the effluent control systems for liquids and solids, the following conclusions may be presented in the technical evaluation report and environmental assessment.

NRC has completed its review of the effluent control systems for liquids and solids proposed for use at the _____ *in situ* leach facility. This review included an evaluation using the review procedures in standard review plan Section 4.2.2 and the acceptance criteria outlined in standard review plan Section 4.2.3.

The applicant has acceptably described the common liquid effluents generated at the facility. Appropriate control methods, including diversion to surface impoundments, deep well injection, and land application/irrigation (select appropriate methods) are identified. On-site evaporation system designs are prescribed in acceptable detail, including engineering plans and drawings. The applicant has shown that liquid waste disposal facilities are adequate to handle production and restoration efforts and has designed installation and operation of surface impoundments such that the impoundments can contain the entire contents of any other leaking or inoperative impoundment. The applicant has described how any dikes used to form a surface impoundment are designed, constructed, and maintained with sufficient structural integrity to prevent massive failure. Additionally, surface impoundments and associated liners are properly designed. The applicant has proposed daily checks of impoundment freeboard and leak

Effluent Control Systems

Table 4.2.3-1. Non-NRC Permits That May Be Required to Support Liquid Effluent Disposal at Uranium *in Situ* Leach Facilities	
Permit	**Comments**
Underground Injection Control	Mandatory. Issued either by EPA or a state under EPA authority. EPA reserves exclusive aquifer exemption action.
Surface-Water Discharge	Optional. Usually issued by the state, under U.S. Environmental Protection Agency (EPA) authority.
Air	Mandatory with dryer. Usually issued by state under EPA authority; may also be local.
Mining	Mandatory. Usually issued by state under legislative authority.
Wetlands	Issued by U.S. Army Corps of Engineers
Consumptive Water Use	Mandatory. Issued by a state under legislative authority. (Secure water rights)
Leases/Permits on Federal Lands	Issued by U.S Bureau of Land Management , U.S. Bureau of Indian Affairs (Department of the Interior), U.S. Forest Services. U.S. Department of Agriculture, or U.S. Bureau of Reclamation.
Construction/Sewage	Issued by local authorities: building codes, utility authorities, and planning authorities.
Leases/Permits on State Lands	Issued by state land offices.

detection systems. Chemical sampling is initiated when levels are greater than 15 cm [6 in.]. The planned sampling and analysis of contaminants in the leak detection systems are acceptable.

An appropriate corrective action plan is described that allows for the contents of a given impoundment to be transferred to another impoundment with no release of contamination. The applicant has an acceptable action plan to notify NRC, analyze samples, and file a written report in the event of leaks. The applicant has ensured that disposal plans are in compliance with applicable directives. Acceptable plans and procedures that address contingencies for all reasonably expected system failures are provided. The applicant has demonstrated that sump capacity is sufficient to contain the volume of the largest hazardous material source. The facility has acceptable alarms to notify the operator of loss of or excess pressure within the production circuits. The applicant log of significant solution spills is acceptable. Applicant plan for spill notification is acceptable. The applicant has an acceptable plan for the disposal of contaminated solid wastes that are generated by the facility. The applicant has proposed storage of contaminated material that either cannot or will not be decontaminated and released for unrestricted use. The applicant has demonstrated that the contamination will be managed to insure compliance with occupational dose limits, as discussed in Section 5.7 of this standard review plan. The applicant has demonstrated possession of the appropriate water quality certification and discharge permits or has plans in place to obtain them. By providing information on the health and safety impacts of system failures and identifying preventive measures and mitigation for such occurrences, the applicant has shown that effluent control systems will limit radiation exposures under both normal and accident conditions. The applicant has committed to maintaining occupational radiation doses and doses to the general public within applicable 10 CFR Part 20 exposure limits and as low as is reasonably achievable.

Based on the information provided in the application and the detailed review conducted of the effluent control systems for liquids and solids for the _____ in situ leach facility, the staff has concluded that the proposed effluent control systems for liquids and solids are acceptable and are in compliance with 10 CFR 20.1101, which requires that an acceptable radiation protection program that achieves as low as is reasonably achievable goals is in place; 10 CFR 20.1201, which defines the allowable occupational dose limits for adults; 10 CFR 20.1301, which defines dose limits allowable for individual members of the public; 10 CFR 20.1302, which requires compliance with dose limits for individual members of the public; 10 CFR 20.2007, which requires that disposal by injection in deep wells must also meet any other applicable federal, state, and local government regulations pertaining to deep well injection; 10 CFR Part 40, Appendix A, Criterion 2, which requires that the applicant provide an estimate of the amount of contaminated material that will be generated and objective evidence of an agreement for disposal of these materials either in a licensed waste disposal site or at a licensed mill tailings facility to demonstrate nonproliferation of waste disposal sites; 10 CFR Part 40, Appendix A, Criteria 5A(1) through 5A(5), which define design provisions for surface impoundments; Criterion 5E which requires measures to protect ground water; Criterion 5F which provides requirements for seepage control; Criterion 5G(1), which requires that the chemical and radioactive characteristics of wastes be defined; Criterion 6(6), which defines cleanup standards for radium. The related reviews of the 10 CFR Part 20 radiological aspects of the effluent control systems for liquids and solid radionuclides, in accordance with standard review plan Sections 5.0, "Operations" and 7.0, "Environmental Effects" are addressed elsewhere in this technical evaluation report.

The design of dikes used to construct surface-water impoundments complies with Regulatory Guide 3.11, Sections 2 and 3 (NRC, 1977), and therefore meet the requirements of 10 CFR Part 40, Appendix A, Criterion 5(A)5. In addition, because the impoundment dikes may meet the definition of a dam as given in the Federal Guidelines for Dam Safety, they are subject to the NRC Dam Safety Program, and to Section 215, "National Dam Safety Program, of the Water Resources Development Act of 1966" (optional, staff should add only if appropriate).

The staff has also considered the environmental impacts from the proposed liquid waste management approach. Considered in the evaluation were the potential environmental impacts as well as alternatives and mitigative measures. In evaluating the environmental impacts, the staff examined effects from radiological as well as non-radiological aspects. Alternatives considered include [staff should list as appropriate]. In addition, the applicant will take the following preventive and mitigative measures to reduce the environmental impacts (staff should

Effluent Control Systems

list measures and discuss how they reduce impact based on this evaluation). The staff has determined that the environmental impacts from the proposed facility are acceptable.

4.2.5 References

NRC. Regulatory Guide 3.11.1 "Operational Inspection and Surveillance of Embankment Retention Systems for Uranium Mill Tailings." Revision 1. Washington, DC: NRC. 1980.

———. Regulatory Guide 3.11, "Design, Construction, and Inspection of Embankment Retention Systems for Uranium Mills." Washington, DC: NRC, Office of Standards Development. 1977.

4.3 <u>Contaminated Equipment</u>

The review in this area will be conducted using Section 5.7.6 of this standard review plan.

5.0 OPERATIONS

5.1 Corporate Organization And Administrative Procedures

5.1.1 Areas of Review

The staff should review the detailed description of the applicant's proposed organization and administrative procedures, including a description and/or chart depicting the key positions in the management structure, and the responsibilities and functions of each with respect to development, review, approval, implementation, and adherence to operating procedures, radiation safety programs, environmental and ground-water monitoring programs, quality assurance programs, routine and non-routine maintenance activities, and changes to any of these. These include procedures that evaluate the consequences of a spill or incident/event against 10 CFR Part 20, Subpart M and 10 CFR 40.60 criteria. In addition, the reviewer should examine the plans proposed by the applicant for establishing a Safety and Environmental Review Panel, or similarly named panel, including the proposed composition and responsibilities of the Panel.

5.1.2 Review Procedures

The staff should determine whether the proposed organization and administrative procedures are defined in sufficient detail to evaluate the responsibilities and authority of persons in positions responsible for developing, reviewing, approving, implementing, and enforcing the proposed programs related to radiological safety, environmental safety, ground-water protection, quality assurance, and maintenance. In addition, the reviewer should examine the plans proposed by the applicant for establishing a Safety and Environmental Review Panel including the proposed composition and responsibilities of the Panel.

For license renewals and amendment applications, Appendix A to this standard review plan provides guidance for examining facility operations and the approach that should be used in evaluating amendments and renewal applications.

5.1.3 Acceptance Criteria

The corporate organization and administrative procedures are acceptable if they meet the following criteria:

(1)   The applicant has provided adequate descriptions of the corporate organization, clearly defining management responsibilities and authority at each level.

Specifically, the radiation safety officer should have the responsibilities and authority outlined in Regulatory Guide 8.31, Section 1.2 (NRC, 2002).

(2)   The organizational structure shows integration among groups that support the operation and maintenance of the facility. If the facility is new, integration between plant construction and plant management should be detailed.

Operations

(3)   The applicant has established a Safety and Environmental Review Panel that will consist of at least three individuals. One member of the Safety and Environmental Review Panel will have expertise in management and will be responsible for implementing managerial and financial changes. One member will have expertise in operations and/or construction and will have responsibility for implementing any operational changes. One member will be the radiation safety officer, or equivalent, with the responsibility for assuring that changes conform to radiation safety and environmental requirements. Additional members may be included in the Safety and Environmental Review Panel, as appropriate, to address specific technical issues such as health physics, ground-water hydrology, surface-water hydrology, and specific earth sciences or other technical disciplines. Temporary members may include consultants. A description of when additional members will be used is provided.

(4)   To the extent possible, proposed administrative procedures conform with Regulatory Guide 8.2, "Guide for Administrative Practices in Radiation Monitoring" (NRC, 1973) and with Regulatory Guide 4.15, "Quality Assurance for Radiological Monitoring Programs (Normal Operations)—Effluent Streams and the Environment, Revision 1, (NRC, 1979).

(5)   Sufficient independence is available to the plant supervisor, radiation safety officer, and Safety and Environmental Review Panel such that significant safety issues can be raised to senior management.

5.1.4   Evaluation Findings

If the staff review, as described in this section, results in the acceptance of the corporate organization and administrative procedures, the following conclusions may be presented in the technical evaluation report.

NRC has completed its review of the corporate organization and administrative procedures proposed for use at the _____ *in situ* leach facility. This review included an evaluation using the review procedures in standard review plan Section 5.1.2 and the acceptance criteria outlined in standard review plan Section 5.1.3.

The applicant has an acceptable corporate organization that defines management responsibilities and authority at each level. The applicant's definition of the responsibilities and procedures with respect to development, review, approval, implementation, and adherence to operating procedures, radiation safety programs, environmental and ground-water monitoring programs, quality assurance programs, routine/non-routine maintenance activities, and changes to any of these is acceptable. Integration among groups that support operation and maintenance of the facility is demonstrated. In the case of a new facility, integration between facility construction and plant management is acceptably detailed. The applicant has established a Safety and Environmental Review Panel with at least three individuals representing expertise in management/financial, operations/construction, and radiation safety matters. The applicant has demonstrated that specific technical issues will be dealt with by the Safety and Environmental Review Panel, with support from other qualified staff members, or consultants, as appropriate.

Based on the information provided in the application and the detailed review conducted of the corporate organization and administrative procedures for the _____ *in situ* leach facility, the staff concludes that the proposed corporate organization and administrative procedures are acceptable and are in compliance with 10 CFR 20.1101, which defines radiation protection program requirements. In addition, the requirements of 10 CFR 40.32(b), (c), and (d) are also met as they relate to the proposed corporate organization and Safety and Environmental Review Panel functions.

5.1.5 References

NRC. Regulatory Guide 8.31, "Information Relevant to Ensuring that Occupational Radiation Exposures at Uranium Mills Will Be As Low As Is Reasonably Achievable." Rev. 1. Washington, DC: NRC, Office of Nuclear Regulatory Research. 2002.

———. Regulatory Guide 4.15, "Quality Assurance for Radiological Monitoring Programs (Normal Operations)–Effluent Streams and the Environment." Revision 1. Washington, DC: NRC, Office of Standards Development. 1979.

———. Regulatory Guide 8.2, "Guide for Administrative Practices in Radiation Monitoring." Washington, DC: NRC, Office of Standards Development. 1973.

5.2 Management Control Program

5.2.1 Areas of Review

The staff should review the management control program and administrative procedures proposed to ensure that activities affecting health, safety, and the environment will be conducted in accordance with written standard operating procedures, including records keeping and reporting. The reviewer should evaluate the management control and decision bases to be used by the Safety and Environmental Review Panel in deciding when it is necessary to apply for a license amendment. Procedures governing non-routine work or maintenance that is not covered by a standard operating procedure, such as use of radiation work permits, should be reviewed.

The staff should examine the applicant's program for cultural resources protection.

The staff should review the applicant's record keeping and retention plans for the materials control and tracking program; the radiation protection program; the sampling, survey and calibration programs; for planned special exposures; to track doses to workers and members of the public; for the disposal of source, and byproduct materials made under 10 CFR 20.2002 and 20.2003; and for the records important to decommissioning the facility, including records of spills or unusual occurrences involving the spread of contamination, cleanup actions taken, and the location of remaining contamination. The staff should also review the licensee's plans and arrangements to identify and maintain the records that must be retained for the life of the facility and ultimately be transferred to NRC at the termination of the license.

Operations

While occupational and safety concerns are important and need to be included in the development of standard operating procedures, NRC regulatory authority is limited to those instances where occupational safety concerns may affect radiological operations or accidents.

5.2.2 Review Procedures

The reviewer should determine that the proposed management control program and administrative procedures are sufficient to assure that any activities affecting health, safety, and the environment, including compliance with any license commitments or conditions, will be conducted in accordance with written operating procedures. The review should include the process for identifying and developing standard operating procedures for routine work, and the review and approval process to be used by the radiation safety staff to modify standard operating procedures when appropriate. Methods for review and approval of non-routine work or maintenance activity by the radiation safety staff should be examined.

The reviewer should determine whether the licensee has agreed to administer a cultural resources inventory before engaging in any development activity not previously assessed by NRC. The reviewer should verify that any disturbances to be associated with such development will be completed in compliance with the National Historic Preservation Act, the Archeological Resources Protection Act, and their implementing regulations. Additionally, the reviewer should evaluate if the licensee has committed to cease any work resulting in the discovery of previously unknown cultural artifacts to ensure that no unapproved disturbance occurs. The reviewer should confirm that any such artifacts will be inventoried and evaluated, and no further disturbance will occur until the licensee has received authorization from the NRC to proceed.

The reviewer should determine whether the proposed record keeping and retention programs are adequate to ensure that the licensee will be able to track, control, and demonstrate control of, the source and byproduct material at the site, such that on-site and off-site dose limits will not be exceeded. The reviewer should determine whether records important to decommissioning, such as descriptions of spills and other unusual occurrences, will be maintained by the licensee, and will be in an identifiable or, preferably, separate file. The reviewer should also determine whether the licensee has a plan to maintain the records that will be turned over to NRC at license termination.

For license renewals and amendment applications, Appendix A to this standard review plan provides guidance for examining facility operations and the approach that should be used in evaluating amendments and renewal applications.

5.2.3 Acceptance Criteria

The management control program is acceptable if

(1)   The proposed management control program is sufficient to assure that all proposed activities that may affect health, safety, and the environment, including compliance with any license commitments or conditions, will be conducted in accordance with written

operating procedures. These shall include procedures that evaluate the consequences of a spill or incident/event against 10 CFR Part 20, Subpart M and 10 CFR 40.60 reporting criteria.

(2)   The applicant provides a process that will be used to identify and prepare operating procedures for routine work.

        There is an adequate mechanism for the development, approval, and review (on an annual basis) of standard operating procedures by the radiation safety staff. Subsequent inspections will ensure that standard operating procedures are adequate and applied correctly.

        The process includes procedures covering all aspects of radiation safety, routine maintenance activities (especially in radiation areas), and Safety and Environmental Review Panel reviews and activities.

        For standard operating procedures for radiation safety, refer to Regulatory Guide 8.31, Section 2 (NRC, 2002).

(3)   The applicant presents methods for review and approval of non-routine work or maintenance activity by the radiation safety staff. The methods include the preparation and issuance of radiation work permits for activities where standard operating procedures do not apply.

(4)   The applicant provides for the establishment of a Safety and Environmental Review Panel. (A detailed review of Safety and Environmental Review Panel composition is addressed in Section 5.1 of this standard review plan.) Procedures governing the functioning of the Safety and Environmental Review Panel ensure that approvals of any changes in the facility, the operating procedures, or the conduct of tests or experiments are appropriately documented and reported. These changes, tests, or experiments may be effected without obtaining a license amendment pursuant to 10 CFR 40.44, so long as the change, test, or experiment does not

        (a)   Create a possibility for an accident of a different type than previously evaluated in the license application (as updated)

        (b)   Create a possibility for a malfunction of a structure, system, or control with a different result than previously evaluated in the license application (as updated)

        (c)   Result in a departure from the method of evaluation described in the license application (as updated) used in establishing the final safety evaluation report or the environmental assessment or technical evaluation reports or other analyses and evaluations for license amendments

        Quantitative likelihood and consequence analyses may not be required for changes at uranium *in situ* leach facilities.

Operations

The Safety and Environmental Review Panel records will include written safety and environmental evaluations made by the Safety and Environmental Review Panel that provide the basis for determining whether changes, tests, or experiments were implemented in accordance with the bases described in Section 5.2.3. Changes pages should have both a change indicator for the area changed (e.g., a bold line vertically drawn in the right margin adjacent to the portion actually changed) and a page change indication (date of change or change number, or both).

The annual Safety and Environmental Review Panel report and page changes may be furnished along with reports normally submitted to satisfy 10 CFR 40.65 reporting requirements.

(5)   The licensee is exempted from the requirements of 20 CFR 1902(e) for areas within the facility, provided that all entrances to the facility are conspicuously posted with the words "ANY AREA WITHIN THIS FACILITY MAY CONTAIN RADIOACTIVE MATERIAL."

(6)   The licensee has agreed to administer a cultural resources inventory before engaging in any development activity not previously assessed by NRC. Any disturbances to be associated with such development will be completed in compliance with the National Historic Preservation Act, the Archeological Resources Protection Act, and their implementing regulations. The licensee has committed to cease any work resulting in the discovery of previously unknown cultural artifacts to ensure that no unapproved disturbance occurs. Any such artifacts will be inventoried and evaluated, and no further disturbance will occur until the licensee has received authorization from the NRC to proceed.

(7)   The record keeping and retention plans demonstrate that the licensee will maintain and retain records of the receipt, transfer, and disposal of any source or byproduct material processed or produced at the licensed facility, for the period set out in the license conditions, or until the Commission terminates the license.

(8)   The following will be permanently maintained and retained until license termination:

(a)   Records of on-site radioactive waste disposal such as by deep well injection, land application, or burial under 10 CFR 20.2002 and 20.2007.

(b)   Records required by 10 CFR 20.2103(b)(4).

(c)   Records required by 10 CFR Part 40, Appendix A, Criteria 8 and 8A and included in Regulatory Guide 3.11.1 (NRC, 1980).

(d)   Records containing information important to decommissioning and reclamation, including

(i)   Descriptions of any spills, excursions, contamination events or unusual occurrences, including the dates, locations, areas, or facilities affected; assessments of hazards; corrective and cleanup actions taken;

assessment of cleanup effectiveness, and the location of any remaining contamination; nuclides involved; quantities, forms and concentrations, and descriptions of hazardous constituents; descriptions of inaccessible areas that cannot be cleaned up; and sketches, diagrams, or drawings marked to show areas of contamination and places where measurements were made. Significant spills that should be included are any radiological spills that have the potential to exceed site cleanup standards and any radiological spill that leaves the site. A license condition will be established to this effect.

(ii)   Information related to site characterization; residual soil contamination levels; on-site locations used for burials of radioactive materials; hydrology and geology, with particular emphasis on conditions that could contribute to ground-water or surface-water contamination; and locations of surface impoundments, waste water ponds, lagoons, and well field aquifer anomalies.

(iii)   As-built drawings or photographs of structures, equipment, restricted areas, well fields, areas where radioactive materials are stored, and any modifications showing the locations of these structures and systems through time.

(iv)   Drawings of areas of possible inaccessible contamination, including features such as buried pipes or pipelines.

(v)   Pre-operational background radiation levels at and near the site.

These records will be maintained with adequate safeguards against tampering and loss.

(9)   The licensee demonstrates that records can be provided to a new owner or new licensee in the event that the property or license is transferred, or to NRC, after license termination.

(10)   New licensees or owners demonstrate that any such records received from a previous owner or licensee will be retained, along with their own records, to be turned over to NRC after license termination.

(11)   Records will be maintained as hard copy originals, as copies on microfiche, or will be electronically protected, and will be readily retrievable for NRC inspection.

(12)   Reports of spills; evaporation pond leaks; excursions of source, 11e.(2) byproduct material, or process chemicals; will be made to the Headquarters Project Manager by telephone or electronic mail (email) within 48 hours of the event. This notification shall be followed, within thirty (30) days of the notification, by submittal of a written report to the NRC Headquarters Project Manager, detailing the conditions leading to the spill or incident/event, corrective actions taken, and results achieved. A license condition will be established to this effect.

Operations

(13)   An annual report will be submitted to the NRC that includes the as low as is reasonably achievable audit report, land use survey, monitoring data, corrective action program report, one of the semiannual effluent and environmental monitoring reports, and the Safety and Environmental Review Panel information. A license condition will be established to this effect.

5.2.4   Evaluation Findings

If the staff review, as described in this section, results in the acceptance of the management control program, the following conclusions may be presented in the technical evaluation report.

NRC has completed its review of the management control program proposed for use at the _____in situ leach facility. This review included an evaluation using the review procedures in standard review plan Section 5.2.2 and the acceptance criteria outlined in standard review plan Section 5.2.3.

The applicant has an acceptable management control program that assures that all safety-related operating activities can be conducted according to written operating procedures. The applicant has provided acceptable operating procedures or a process that will be used to develop standard operating procedures. The applicant has acceptably identified radiation protection, maintenance activities (especially in radiation areas), development of well fields, and Safety and Environmental Review Panel reviews as areas where standard operating procedures are acceptable and correctly applied. The applicant has demonstrated that non-routine work or maintenance activity will comply with radiation safety requirements and that radiation work permits will be issued for activities where standard operating procedures do not apply.

The applicant will administer a cultural resources protection program in compliance with the National Historic Preservation Act, the Archeological Resources Protection Act, and their implementing regulations. The applicant will cease any work resulting in the discovery of previously unknown cultural artifacts until such artifacts are inventoried and evaluated and authorization has been obtained from the NRC to proceed.

The applicant has acceptable record keeping and retention and reporting programs that will be adequate to ensure that the licensee is able to track, control, and demonstrate control over the source and byproduct materials that are processed, produced, or stored at the facility during its operating life, through decommissioning, and to license termination. The record keeping and retention plans will assist in ensuring that both on-site and off-site exposures are kept within regulatory limits and in documenting compliance with NRC regulations. The applicant has demonstrated an acceptable program to maintain records on spills, likely contamination events, and unusual occurrences for use in calculating annual surety amounts and to ensure acceptable decommissioning. The applicant will maintain records for decommissioning, on-site and off-site disposal, personnel exposure, and off-site releases of radioactivity, as permanent records for the facility that will be transferred to any new owner or licensee, and ultimately to NRC, before license termination. Reports will be made to the NRC as required by regulations.

Based on the information provided in the application and the detailed review conducted of the management control program for the _____*in situ* leach facility, the staff concludes that the proposed management control program is acceptable and is in compliance with 10 CFR Part 40, Appendix A, Criteria 8 and 8A, which specify documentation requirements for airborne effluents and waste retention systems; 10 CFR 20.1101, which defines radiation protection program requirements; the National Historic Preservation Act and the Archeological Resources Protection Act, which define requirements for the protection of cultural resources; 10 CFR Part 20, Subpart L and Subpart M, which define requirements for record keeping and reporting; and 10 CFR 40.61(d) and (e), which also define requirements for record keeping.

5.2.5 References

NRC. Regulatory Guide 8.31, "Information Relevant to Ensuring that Occupational Radiation Exposures at Uranium Mills will be as low as is Reasonably Achievable." Rev. 1. Washington, DC: NRC, Office of Nuclear Regulatory Research. 2002.

—— Regulatory Guide 3.11.1, Revision 1, "Operational Inspection and Surveillance of Embankment Retention Systems for Uranium Mills." Washington, DC: NRC, Office of Standards Development. 1980.

5.3   Management Audit and Inspection Program

5.3.1 Areas of Review

The staff should review the proposed management audit, inspection, and as low as is reasonably achievable program, including the frequencies, types, and scopes of reviews and inspections; action levels; corrective action measures; and the responsibilities of each participant. The staff should also review the program for ensuring that employee exposures (to both airborne and external radiation) and effluent releases are as low as is reasonably achievable.

5.3.2 Review Procedures

The reviewer should determine whether the management audit and inspection program is acceptable and will provide reasonable assurance that employee exposures and effluent releases will be as low as is reasonably achievable. The reviewer shall ensure that yellowcake drying and packaging operations are in accordance with 10 CFR Part 40, Appendix A, Criterion 8, and inspection of waste retention systems is in accordance with Criterion 8A.

For license renewals and amendment applications, Appendix A to this standard review plan provides guidance for examining facility operations and the approach that should be used in evaluating amendments and renewal applications.

Operations

5.3.3 Acceptance Criteria

The management audit, and inspection program is acceptable if it meets the following criteria:

(1)   The proposed frequencies, types, and scopes of reviews and inspections; action levels; and corrective action measures are acceptable to implement the proposed controls.

Management responsibilities for audit and inspection are adequately defined. Acceptable programs for inspection of embankment systems on a regular basis are described in Regulatory Guide 3.11 (NRC, 1977) and Regulatory Guide 3.11.1 (NRC, 1980).

Acceptable programs for annual as low as is reasonably achievable audits are described in Regulatory Guide 8.31 (NRC, 2002).

5.3.4 Evaluation Findings

If the staff review, as described in this section, results in the acceptance of the management audit and inspection program, the following conclusions may be presented in the technical evaluation report.

NRC has completed its review of the management audit and inspection program proposed for use at the _____ *in situ* leach facility. This review included an evaluation using the review procedures in standard review plan Section 5.3.2 and the acceptance criteria outlined in standard review plan Section 5.3.3.

The applicant has an acceptable management audit and inspection program that provides frequencies, types, and scopes of reviews and inspections; action levels; and corrective action measures sufficient to implement the proposed actions.

Based on the information provided in the application and the detailed review conducted of the management audit and inspection program for the _____ *in situ* leach facility, the staff concludes that the proposed programs are acceptable and are in compliance with 10 CFR 20.1702, which requires the use of process or other engineering measures to control the concentrations of radioactive material in the air; and 10 CFR 20.1101 which contains requirements for maintaining radiation exposure limits as low as is reasonably achievable. In addition, the requirements of 10 CFR 40.32(b), (c), and (d) are met as they relate to the acceptability of management audits to ensure protection of health and minimize danger to life and property. The requirements of 10 CFR Part 40, Appendix A, Criteria 8 and 8A are met as they relate to yellowcake drying and packaging operations, and inspection of waste retention systems.

5.3.5 References

NRC. Regulatory Guide 8.31, "Information Relevant to Ensuring that Occupational Radiation Exposures at Uranium Mills will be as low as is Reasonably Achievable." Rev. 1. Washington, DC: NRC, Office of Nuclear Regulatory Research. 2002.

—— Regulatory Guide 3.11.1, Revision 1, "Operational Inspection and Surveillance of Embankment Retention Systems for Uranium Mills." Washington, DC: NRC, Office of Standards Development. 1980.

———. Regulatory Guide 3.11, "Design, Construction, and Inspection of Embankment Retention Systems for Uranium Mills." Revision 2. Washington, DC: NRC, Office of Standards Development. 1977.

5.4   Qualifications for Personnel Conducting the Radiation Safety Program

5.4.1   Areas of Review

The staff should review descriptions of the minimum qualifications and experience levels required for personnel who will be assigned the responsibility for developing, conducting, and administering the radiation safety program. The staff should also review the qualifications of people specifically proposed for these positions.

5.4.2   Review Procedures

The reviewer should determine whether the minimum qualifications and experience levels required for personnel who will be assigned the responsibility for developing, conducting, and administering the radiation safety program are sufficient to meet the guidance provided by Regulatory Guide 8.31 (NRC, 2002). The staff should also determine whether the qualifications of people specifically proposed for these positions are consistent with the minimum qualifications and experience levels.

For license renewals and amendment applications, Appendix A to this standard review plan provides guidance for examining facility operations and the approach that should be used in evaluating amendments and renewal applications.

5.4.3   Acceptance Criteria

The qualifications of radiation safety personnel are acceptable if they meet the following criteria:

(1)   The personnel meet minimum qualifications and experience for radiation safety staff that are consistent with Regulatory Guide 8.31, Section 2.4 (NRC, 2002). The emphasis of this guidance is for uranium recovery facilities; however, the training requirements apply equally to *in situ* leach facilities.

5.4.4   Evaluation Findings

If the staff review, as described in this section, results in the acceptance of the qualifications of facility personnel conducting the radiation safety program, the following conclusions may be presented in the technical evaluation report.

Operations

NRC has completed its review of the qualifications of facility personnel conducting the radiation safety program at the _____ *in situ* leach facility. This review included an evaluation using the review procedures in standard review plan Section 5.4.2 and the acceptance criteria outlined in standard review plan Section 5.4.3.

Based on the information provided in the application and the detailed review conducted of the qualifications of the personnel conducting the radiation safety program for the_____ *in situ* leach facility, the staff concludes that the qualifications of the personnel are acceptable and are in compliance with 10 CFR 20.1101, which defines radiation protection program requirements, and 10 CFR 40.32(b), which provides requirements for applicant qualifications. The qualifications of personnel conducting the radiation safety program are acceptable consistent with NRC Regulatory Guide 8.31 (NRC, 2002).

5.4.5 Reference

NRC. Regulatory Guide 8.31, "Information Relevant to Ensuring that Occupational Radiation Exposures at Uranium Mills Will Be As Low As Is Reasonably Achievable." Rev. 1. Washington, DC: NRC, Office of Nuclear Regulatory Research. 2002.

5.5   Radiation Safety Training

5.5.1   Areas of Review

The staff should review the proposed radiation safety training program, including the content of the initial training or indoctrination, testing, on-the-job training, and the extent and frequency of retraining. The staff should also review the proposed written radiological safety instructions that will be provided to employees to include personal hygiene, contamination surveying before eating or leaving the operating area, requirements for personal monitoring devices and respirators, house keeping requirements, spill cleanup procedures, and emergency actions.

5.5.2   Review Procedures

The staff will examine plans for initial training or indoctrination, testing, on-the-job training, and the extent and frequency of retraining to determine whether they are consistent with Regulatory Guide 8.31 (NRC 2002), Regulatory Guide 8.13 (NRC, 1999), and Regulatory Guide 8.29 (NRC, 1996).

The staff should determine whether the applicant has a radiation safety training program that is adequate to provide radiological safety instructions to the employees. The staff should also determine whether the proposed radiological safety instructions that will be provided to employees will be sufficiently detailed to meet acceptance criteria identified in Section 5.5.3.

For license renewals and amendment applications, Appendix A to this standard review plan provides guidance for examining facility operations and the approach that should be used in evaluating amendments and renewal applications.

5.5.3 Acceptance Criteria

The training program is acceptable if it meets the following criteria:

(1)   It is consistent with the approach described in Regulatory Guide 8.31, Section 2.5 (NRC, 2002).

This guide recommends that before beginning their jobs, all new employees should be instructed, by means of an established course, in the inherent risks of exposure to radiation and the fundamentals of protection against exposure to uranium and its daughters.

(2)   It is consistent with Regulatory Guide 8.13, "Instruction Concerning Prenatal Radiation Exposure, Revision 3" (NRC, 1999).

This guide provides guidance for protection of the fetus.

(3)   It is consistent with Regulatory Guide 8.29, "Instruction Concerning Risks from Occupational Radiation Exposure, Revision 1" (NRC, 1996).

This guide provides a basis for training employees on the risks from radiation exposure in the work place.

5.5.4 Evaluation Findings

If the staff review, as described in this section, results in the acceptance of the radiation safety training program, the following conclusions may be presented in the technical evaluation report.

NRC has completed its review of the radiation safety training program at the _____*in situ* leach facility. This review included an evaluation using the review procedures in standard review plan Section 5.5.2 and the acceptance criteria outlined in standard review plan Section 5.5.3.

The radiation safety training program at the _____*in situ* leach site is consistent with the guidance contained in NRC Regulatory Guides 8.31 (NRC, 2002), 8.13 (NRC, 1999), and 8.29 (NRC, 1996). The content of the training material, testing, on-the-job training, and the extent and frequency of retraining are acceptable. Radiation safety instructions for employees are acceptable.

Based on the information provided in the application and the detailed review conducted of the radiation safety training program for the _____ *in situ* leach facility, the staff concludes that the radiation safety training program is acceptable and is in compliance with 10 CFR 20.1101, which defines radiation protection program requirements, and 10 CFR 40.32(b), as it relates to applicant qualifications through training.

Operations

5.5.5 References

NRC. Regulatory Guide 8.31, "Information Relevant to Ensuring That Occupational Radiation Exposures at Uranium Mills Will Be As Low As Is Reasonably Achievable." Rev. 1. Washington, DC: NRC, Office of Nuclear Regulatory Research. 2002.

————. Regulatory Guide 8.13, "Instruction Concerning Prenatal Radiation Exposure." Revision 3. Washington, DC: NRC, Office of Standards Development. 1999.

————. Regulatory Guide 8.29, "Instruction Concerning Risks from Occupational Radiation Exposure." Revision 1." Washington, DC: NRC, Office of Standards Development. 1996.

5.6   Security

5.6.1   Areas of Review

The staff should review the security measures proposed to prevent unauthorized entry into the controlled area.

5.6.2   Review Procedures

The staff should determine whether the proposed security measures are sufficient to prevent unauthorized entry into the controlled area in accordance with regulatory requirements in 10 CFR Part 20, Subpart I.

For license renewals and amendment applications, Appendix A to this standard review plan provides guidance for examining facility operations and the approach that should be used in evaluating amendments and renewal applications.

5.6.3   Acceptance Criteria

The security program is acceptable if the applicant has acceptable passive controls, such as fencing for well fields, and active controls, such as daily inspections and locks for plant buildings.

5.6.4   Evaluation Findings

If the staff review, as described in this section, results in the acceptance of the security measures, the following conclusions may be presented in the technical evaluation report and environmental assessment.

NRC has completed its review of the security measures at the _____ *in situ* leach facility. This review included an evaluation using the review procedures in standard review plan Section 5.6.2 and the acceptance criteria outlined in standard review plan Section 5.6.3.

The security measures at the _____ in situ leach site demonstrate that the applicant has acceptable active and passive constraints on entry to the licensed and restricted areas. The applicant has identified acceptable passive controls, for example, barbed wire fencing, locked gates, and warning signage for site control and active security systems for buildings.

Based on the information provided in the application and the detailed review conducted of the security measures for the _____ in situ leach facility, the staff concludes that the security measures are acceptable and are in compliance with 10 CFR Part 20, Subpart I, which provides requirements for the security of stored material and control of material not in storage.

5.6.5 References

None.

5.7 Radiation Safety Controls And Monitoring

5.7.1 Effluent Control Techniques

5.7.1.1 Areas of Review

The staff should review descriptions of the effluent control techniques (e.g., ventilation, confinement, filtration) designed to minimize in-plant and environmental emissions at each step of the process where releases might occur. Major airborne radioactive effluents include radioactive particulates (from drying and packaging areas) and radon gas emanating from production solutions. Radon gas mobilization can occur from recovery solutions at process locations where systems allow venting. The staff should evaluate effluent control techniques for uranium particulate emissions located in drying and packaging areas and in any other areas where release of significant quantities of uranium particulate is a concern. Closed systems can eliminate releases of uranium particulates and radon gas. For example, the use of vacuum packaging equipment has been shown to eliminate uranium releases from packaging operations.

Common liquid effluent sources are process bleed, process solutions (e.g., backwash, resin transfer waters), and wash-down water. The staff should review the facility design for containment of contamination from spills resulting from normal operations and probable accidents (e.g., tank, valve, or pipe joint failure). For surface impoundments used in the management of 11e.(2) byproduct material, the staff should also review engineering design to ensure proper containment performance, and evaluate leak detection and monitoring systems for surface impoundments containing contaminated effluents.

The staff reviews should include minimum performance specifications such as filtration or scrubber efficiency and ventilation airflow at their reasonably expected best performance and the frequency of tests and inspections to ensure that these specifications are being met.

The staff should review contingency plans and notification requirements to be implemented in the event of equipment failures, spills, or excursions.

Operations

5.7.1.2   Review Procedures

The staff should determine whether the proposed effluent control techniques are sufficient to limit radiation exposures and radioactive releases to as low as is reasonably achievable and to ensure conformance with regulatory requirements identified in 10 CFR Part 20.

In general, the reviewer should be familiar with 10 CFR Part 40, Appendix A, Criterion 8 and Regulatory Guide 8.10, "Operating Philosophy for Maintaining Occupational Radiation Exposures As Low As Is Reasonably Achievable" (NRC, 1977). Additional guidance is found in Regulatory Guide 8.37, "ALARA Levels for Effluent from Materials Facilities" (NRC, 1993); Regulatory Guide 8.31, "Information Relevant to Ensuring that Occupational Radiation Exposures at Uranium Mill Will Be As Low As Is Reasonably Achievable" (NRC, 2002); and Regulatory Guide 3.56, "General Guidance for Designing, Testing, Operating, and Maintaining Emission Control Devices at Uranium Mills" (NRC, 1986). The staff should determine whether the proposed effluent control techniques (e.g., ventilation, confinement, filtration) are acceptably described and sufficient to control in-plant and environmental emissions at each step of the process where releases might occur. The staff should ensure that minimum performance specifications for ventilation, filtration, and confinement systems throughout the recovery plant and laboratories are provided and are consistent with assumptions made in exposure estimates for areas of the facility where the systems are operating. The staff should also check that the frequencies of equipment tests and inspections are consistent with manufacturers' recommendations to ensure that these specifications are being met. Engineering design should be adequate to meet the performance specifications. Contingencies for equipment failures, maintenance shutdowns, and spills should be reviewed to ensure procedures are in place to maintain exposures as low as is reasonably achievable.

For license renewals and amendment applications, Appendix A to this standard review plan provides guidance for examining facility operations and the approach that should be used in evaluating amendments and renewal applications.

5.7.1.3   Acceptance Criteria

The effluent control techniques are acceptable if they meet the following criteria:

(1)   Radon gas from processing tanks within enclosed buildings is properly controlled.

Effective control of radon gas can be achieved by using a pressurized processing tank system that eliminates venting in process buildings, or by using appropriate ventilation systems in buildings where radon gas venting is expected.

(2)   Emissions from yellowcake drying operations are properly controlled.

Acceptable control of yellowcake emissions from the dryer is achieved by meeting the criteria of 10 CFR Part 40, Appendix A, Criterion 8 and Regulatory Guide 3.56, Section 1 (NRC, 1986).

(3)   Release of liquids into surface waters must comply with the public dose limits in 10 CFR 20.1301, which may be demonstrated by one of the following methods:

   (a) The licensee demonstrates compliance with 10 CFR Part 20, Appendix B, by one of the following methods and shows that if an individual were continuously present in an unrestricted area, the dose from external sources would not exceed 0.02 mSv/hr [2 mrem/hr] or 0.5 mSv/yr [50 mrem/yr]:

      (i) Showing that the discharge of effluent from any surface impoundment is within 10 CFR Part 20, Appendix B, limits at the point of discharge.

      (ii) Monitoring the incoming process water to demonstrate compliance with the effluent discharge requirements of 10 CFR Part 20, Appendix B, for process water.

   (b) The licensee demonstrates that the total effective dose equivalent to the individual likely to receive the highest dose from the facility does not exceed the annual dose limit for the public.

(4)   The applicant describes minimum performance specifications for the operation of the effluent controls and the frequencies of tests and inspections to ensure proper performance to specifications. Details of acceptable excursion control techniques are found in Section 5.7.8.3 of this standard review plan.

   Acceptable methods for testing, maintenance, and inspection of effluent controls are given in Regulatory Guide 3.56, Section 1 (NRC, 1986).

(5)   Record keeping for the effluent control techniques is sufficient to meet requirements in 10 CFR 20.2103(b)(4).

(6)   The applicant describes emergency procedures in the event of equipment failures or spills, references existing emergency procedures, or commits to the development of emergency procedures.

   For license renewal applications, the historical effluent control program summary is included through the most recent reporting period preceding the submittal of the application.

   The effectiveness of the historical program should be discussed with regard to all applicable 10 CFR Part 20 regulatory requirements identified in the preceding paragraphs. Long-term trends should be discussed, and any short-term deviations from the long-term trend should be explained.

(7)   The effluent control techniques are designed to keep exposures to members of the public as low as is reasonably achievable as described in Regulatory Guide 8.37, Section 2 (NRC, 1993).

Operations

(8)   The effluent control techniques are designed to limit exposures to members of the public from emissions to air (excluding Radon-222 and progeny) to no greater than 0.1 mSv [10 mrem/yr].

5.7.1.4   Evaluation Findings

If the staff review, as described in this section, results in the acceptance of the effluent control techniques, the following conclusions may be presented in the technical evaluation report and environmental assessment.

NRC has completed its review of the effluent control techniques at the _____ *in situ* leach facility. This review included an evaluation using the review procedures in standard review plan Section 5.7.1.2 and the acceptance criteria outlined in standard review plan Section 5.7.1.3.

The applicant has acceptable effluent control techniques at the _____ *in situ* leach site and has demonstrated that important effluent streams are controlled and monitored. The applicant has used an acceptable pressurized processing tank system or appropriate ventilation systems in buildings where radon gas is vented. Acceptable control of the yellowcake dryer system is evidenced by a vacuum dryer or other appropriate particulate scrubber equipment on the dryer stack. The applicant has shown that the discharge of process water is within the dose limits of 10 CFR 20.1301. The applicant has demonstrated acceptable effluent control techniques and associated test and inspection frequencies to ensure specified performance. Record keeping and monitoring procedures are acceptable. Acceptable emergency procedures for managing equipment failures or spills are described by the applicant.

Based on the information provided in the application and the detailed review conducted of the effluent control techniques at the _____ *in situ* leach facility, the staff concludes that this program is acceptable and is in compliance with 10 CFR 20.1301, which provides dose limits for members of the public; 10 CFR 20.1101, which defines radiation protection program and as low as is reasonably achievable requirements; 10 CFR 20.1201(a), which provides occupational dose limits; and 10 CFR Part 20, Subpart M, which defines requirements for reports. In addition, the staff concludes that the effluent control techniques meet the requirements of 10 CFR 40.32(b) to protect health and minimize danger to life and property, and 10 CFR Part 40, Appendix A, Criterion 8, which specifies standards for yellowcake dryer operations.

5.7.1.5   References

NRC. Regulatory Guide 8.31, "Information Relevant to Ensuring that Occupational Radiation Exposures at Uranium Mills Will Be As Low As Is Reasonably Achievable." Rev. 1. Washington, DC: NRC, Office of Nuclear Regulatory Research. 2002.

———. Regulatory Guide 8.37, "ALARA Levels for Effluent from Materials Facilities." Washington, DC: NRC, Office of Standards Development. 1993.

————. Regulatory Guide 3.56, "General Guidance for Designing, Testing, Operating, and Maintaining Emission Control Devices at Uranium Mills." Washington, DC: NRC, Office of Standards Development. 1986.

————. Regulatory Guide 8.10, "Operating Philosophy for Maintaining Occupational Radiation Exposures as low as is Reasonably Achievable." Revision 1–R. Washington, DC: NRC, Office of Standards Development. 1977.

5.7.2 External Radiation Exposure Monitoring Program

5.7.2.1 Areas of Review

The staff should review survey methods, instrumentation, and equipment for determining exposures of employees to external radiation during routine and non-routine operations, maintenance, and cleanup activities. This review should include the types of surveys conducted, criteria for determining survey locations, frequency of surveys, action levels, management audits, and corrective action requirements. Staff should also review the program for personnel exposure monitoring, the criteria for including workers in the program, the sensitivity and range of devices used, and calibration frequency and methods.

5.7.2.2 Review Procedures

The staff should determine whether proposed monitoring methods, instrumentation, and equipment are sufficient to meet the regulatory requirements for determining the exposures of employees to external radiation in 10 CFR 20.1203. In conducting its review, the staff should ensure that the applicant has provided one or more charts that identify the facility layout and the location of monitors for external radiation as well as providing acceptable criteria for determining the sampling locations. The staff should ensure all monitoring equipment will be identified by type with additional specification of the range, sensitivity, calibration methods and frequency, availability, and planned use. Staff should ensure that the proposed monitoring program is sufficient to adequately protect workers from hazards of beta radiation (skin, extremity, lens of eye) resulting from the decay products of U-238 when effective shielding is not present (e.g., maintenance operations). The staff should also ensure that the monitoring program is acceptable to detect and control gamma radiation from uranium decay products in areas where large volumes of uranium may be present (e.g., processing tanks, yellowcake storage areas).

For license renewals and amendment applications, Appendix A to this standard review plan provides guidance for examining facility operations and the approach that should be used in evaluating amendments and renewal applications.

Operations

5.7.2.3   Acceptance Criteria

The external radiation exposure monitoring program is acceptable if it meets the following criteria:

(1)   The application contains one or more drawings that depict the facility layout and the location of monitors for external radiation. Criteria for determining the external radiation monitor locations, are consistent with Regulatory Guide 4.14, Sections 1.1.5 and 2.1.6 (NRC, 1980).

(2)   The application provides criteria to be used in establishing which employees are to receive external exposure monitoring. These criteria are consistent with Regulatory Guide 8.34, "Monitoring Criteria and Methods to Calculate Occupational Radiation Doses," Section C (NRC, 1992a).

(3)   Monitoring equipment is identified by type, sensitivity, calibration methods and frequency, availability, and planned use to protect health and safety. The ranges of sensitivity for the proposed external radiation monitors are consistent with those appropriate to the facility operation.

(4)   All monitoring equipment has a lower limit of detection that allows measurement of 10 percent of the applicable limits. Planned surveys of external radiation are consistent with the guidance in Regulatory Guide 8.30, "Health Physics Surveys in Uranium Mills," Section 1 (NRC, 2002a).

(5)   Plans for documentation of radiation exposures are consistent with the approach in Regulatory Guide 8.7, "Instructions for Recording and Reporting Occupational Radiation Exposure Data, Revision 1" (NRC, 1992b).

(6)   The application presents radiation dose levels for corrective action that are consistent with the 10 CFR Part 20 regulatory requirements.

(7)   Radiation doses will be kept as low as is reasonably achievable by following Regulatory Guide 8.10 (NRC, 1977) and Regulatory Guide 8.31 (NRC, 2002b).

(8)   The applicant monitoring program is adequate to protect workers from hazards of beta radiation (skin, extremity, lens of eye) resulting from the decay products of uranium-238 when effective shielding is not present (e.g., maintenance operations) and is consistent with Regulatory Guide 8.30 (NRC, 2002a).

(9)   The monitoring program is sufficient to detect and control gamma radiation from uranium decay products in areas where large volumes of uranium may be present (e.g., processing tanks, yellowcake storage areas) and is consistent with Regulatory Guide 8.30 (NRC, 2002a).

(10)   The program for external exposure monitoring and determining doses from external exposure is consistent with Regulatory Guide 8.34, Section C (NRC, 1992a).

5.7.2.4   Evaluation Findings

If the staff review, as described in this section, results in the acceptance of the external radiation exposure monitoring program, the following conclusions may be presented in the technical evaluation report.

NRC has completed its review of the external radiation exposure monitoring program at the _____ *in situ* leach facility. This review included an evaluation using the review procedures in standard review plan Section 5.7.2.2 and the acceptance criteria outlined in standard review plan Section 5.7.2.3.

The applicant has proposed an acceptable external radiation exposure monitoring program at the _____ *in situ* leach site. The applicant has provided an acceptable drawing(s) that depicts the facility layout and the location of external radiation monitors. The external radiation monitors are acceptably placed. The applicant has established appropriate criteria to determine which employees should receive external radiation monitoring. The applicant has demonstrated that the range, sensitivity, and calibration of external radiation monitors will protect health and safety of employees during the full scope of facility operations. Planned radiation surveys are adequate. Planned documentation of radiation exposures is acceptable. The applicant's monitoring program is acceptable to protect workers from beta and gamma radiation.

Based on the information provided in the application and the detailed review conducted of the external radiation exposure monitoring program at the _____ *in situ* leach facility, the staff concludes that the external radiation exposure monitoring program is acceptable and is in compliance with 10 CFR 20.1101, which defines a radiation protection program and as low as is reasonably achievable requirements; 10 CFR 20.1201(a), which defines occupational dose limits; 10 CFR 20.1501, which provides requirements of surveying and radiation monitoring; 10 CFR 20.1502, which defines conditions requiring individual monitoring of external dose; 10 CFR Part 20, Subpart L, which specifies record keeping requirements; and 10 CFR Part 20, Subpart M, which defines reporting requirements.

5.7.2.5   References

NRC.   Regulatory Guide 8.30, "Health Physics Surveys in Uranium Recovery Facilities." Washington, DC: NRC, Office of Nuclear Regulatory Research. 2002a.

———. Regulatory Guide 8.31, "Information Relevant to Ensuring That Occupational Radiation Exposures at Uranium Recovery Facilities Will Be As Low As Is Reasonably Achievable." Rev. 1. Washington, DC: NRC, Office of Nuclear Regulatory Research. 2002b.

———. Regulatory Guide 8.34, "Monitoring Criteria and Methods to Calculate Occupational Radiation Doses." Washington, DC: NRC, Office of Standards Development. 1992a.

Operations

———. Regulatory Guide 8.7, "Instructions for Recording and Reporting Occupational Radiation Exposure Data." Revision 1. Washington, DC: NRC, Office of Standards Development. 1992b.

———. Regulatory Guide 4.14, "Radiological Effluent and Environmental Monitoring at Uranium Mills." Washington, DC: NRC, Office of Standards Development. 1980.

———. Regulatory Guide 8.10, "Operating Philosophy for Maintaining Occupational Radiation Exposures As Low As Is Reasonably Achievable." Washington, DC: NRC, Office of Standards Development. 1977.

5.7.3 Airborne Radiation Monitoring Program

5.7.3.1 Areas of Review

The staff should review the proposed airborne radiation monitoring program to determine concentrations of airborne radioactive materials (including radon) during routine and non-routine operations, maintenance, and cleanup. This review should include criteria for determining airborne radiation monitoring locations and sampling frequency with respect to process operations and personnel occupancy, as well as analytical procedures and sensitivity and instrument calibration requirements. Action levels, audits, and corrective action requirements should also be evaluated.

5.7.3.2 Review Procedures

The staff should determine whether the airborne radiation monitoring program proposed by the applicant is sufficient to limit airborne radiation exposures and airborne radioactive releases to as low as is reasonably achievable and is in conformance with regulatory requirements identified in 10 CFR Part 20. The staff should evaluate whether the proposed sampling program to determine concentrations of airborne radioactive materials (including radon) during routine and non-routine operations, maintenance, and cleanup is in conformance with the regulatory requirements identified in 10 CFR 20.1301; 20.1501; 20.1502; 20.1204; and the other applicable requirements listed in Section 5.7.3.3 of this standard review plan. The staff should determine whether action levels, audits, and corrective actions will be consistent with these requirements.

For license renewals and amendment applications, Appendix A to this standard review plan provides guidance for examining facility operations and the approach that should be used in evaluating amendments and renewal applications.

5.7.3.3   Acceptance Criteria

The airborne radiation monitoring program is acceptable if it meets the following criteria:

(1)   The applicant provides one or more drawings that depict the facility layout and the location of samplers for airborne radiation. Locations are based, in part, on a determination of airflow patterns in areas where monitoring is needed, and determination of monitoring locations is consistent with Regulatory Guide 8.30, "Health Physics Surveys in Uranium Recovery Facilities," (NRC, 2002a).

(2)   Monitoring equipment is identified by type, sensitivity, calibration methods and frequency, availability, and planned use to accurately measure concentrations of airborne radioactive species. The application also demonstrates that the ranges of sensitivity are appropriate for the facility operation.

(3)   Planned surveys of airborne radiation are consistent with the guidance in Regulatory Guide 8.30 (NRC, 2002a).

(4)   The proposed monitoring program is sufficient to adequately protect workers from radon gas releases from venting of processing tanks and from yellowcake dust from drying operations, spills, and maintenance activities and is consistent with Regulatory Guide 4.14, Sections 1.1 and 2.1 (NRC, 1980). The air sampling program is consistent with Regulatory Guide 8.30 (NRC, 2002a).

(5)   Plans for documentation of radiation exposures are consistent with the requirements in 10 CFR 20.2102, 20.2103, 20.2106, and 20.2110.

(6)   The applicant demonstrates that respirators will routinely be used for operations within drying and packaging areas and identifies the criteria for determining when respirators will be required for special jobs or emergency situations. The respiratory protection program should be consistent with guidance in Regulatory Guide 8.15, Revision 1, "Acceptable Programs for Respiratory Protection" (NRC, 1999) and Regulatory Guide 8.31, Section 2.7 (NRC, 2002b).

(7)   For license renewal applications, the historical results summary of the airborne radiation monitoring program is included through the most recent reporting period preceding the submittal of the application. The effectiveness of the historical program is discussed with regard to all applicable 10 CFR Part 20 regulatory requirements identified in the preceding paragraphs. Long-term trends are discussed, and any short-term deviations from the long-term trend are explained.

5.7.3.4   Evaluation Findings

If the staff review, as described in this section, results in the acceptance of the airborne radiation monitoring program, the following conclusions may be presented in the technical evaluation report.

Operations

NRC has completed its review of the airborne radiation monitoring program at the _____ *in situ* leach facility. This review included an evaluation using the review procedures in standard review plan Section 5.7.3.2 and the acceptance criteria outlined in standard review plan Section 5.7.3.3.

The applicant has an acceptable airborne radiation monitoring program at the _____ *in situ* leach site. The applicant has provided an acceptable drawing(s) that depicts the facility layout and the locations of airborne radiation monitors. The airborne radiation monitors are acceptably placed. The applicant demonstrated that the range, sensitivity, and calibration of monitors of airborne radiation will enable accurate determinations of the concentrations of airborne radioactive species so as to protect the health and safety of employees during facility operations. The workers are acceptably protected from radon gas releases from venting of processing tanks and from yellowcake dust from drying operations, spills, and maintenance activities. Planned radiation surveys are acceptable. Planned documentation of radiation exposures is consistent with the requirements. The applicant's respiratory protection program is acceptable. The applicant program for monitoring uranium and sampling of radon or its daughters is acceptable. Employee internal exposure calculations will be performed in accordance with 10 CFR 20.1204(a).

Based on the information provided in the application and the detailed review conducted of the airborne radiation monitoring program at the _____ *in situ* leach facility, the staff has concluded that the airborne radiation monitoring program is acceptable and is in compliance with 10 CFR 20.1101, which defines radiation protection program and as low as is reasonably achievable requirements; 10 CFR 20.1201(a), which provides individual occupational dose limits; 10 CFR 20.1201(e), which specifies allowed intake of soluble uranium; 10 CFR 20.1202, which describes the means of compliance when summing internal and external doses; 10 CFR 20.1203, for determination of dose from airborne external radiation; 10 CFR 20.1208, which specifies the exposure limits to a fetus during pregnancy; 10 CFR 20.1301 which identifies public dose limits; 10 CFR 20.1702, which allows employees to limit dose to individuals by controlling access, limiting exposure times, prescribing use of respiratory equipment, or use of other controls; 10 CFR Part 20, Subpart L, which specifies record keeping requirements; 10 CFR Part 20, Subpart M, which provides requirements for reports and notification; and 10 CFR Part 40, Appendix A, Criterion 8, which provides requirements for control of airborne effluents.

5.7.3.5   References

NRC.   Regulatory Guide 8.30, "Health Physics Surveys in Uranium Recovery Facilities." Washington, DC: NRC, Office of Nuclear Regulatory Research. 2002a.

————. Regulatory Guide 8.31, "Information Relevant to Ensuring that Occupational Radiation Exposures at Uranium Mills Will Be As Low As Reasonably Achievable." Rev. 1. Washington, DC: NRC, Office of Nuclear Regulatory Research. 2002b.

————. Regulatory Guide 8.15, "Acceptable Programs for Respiratory Protection." Revision 1. Washington, DC: NRC, Office of Standards Development. 1999.

—————. Regulatory Guide 4.14, "Radiological Effluent and Environmental Monitoring at Uranium Mills." Washington, DC: NRC, Office of Standards Development. 1980.

5.7.4 Exposure Calculations

5.7.4.1   Areas of Review

The staff should review the methodologies proposed to calculate the exposures to radioactive materials by personnel in work areas where airborne radioactive materials could exist. This review should include methods to determine exposures during routine and non-routine operations, maintenance, and cleanup activities.

5.7.4.2   Review Procedures

The staff should evaluate whether the methodologies proposed to calculate the intake of radioactive materials by personnel in work areas where airborne radioactive materials could exist are in accordance with 10 CFR 20.1204 and 20.1201. The review should also place emphasis on the parameters used in exposure calculations to ensure they are representative of conditions at the site. Estimation of airborne uranium concentrations should take into account the maximum production capacity requested in the application and the anticipated efficiencies of airborne particulate control systems reviewed using Section 5.7.1 of this standard review plan.

For license renewals and amendment applications, Appendix A to this standard review plan provides guidance for examining facility operations and the approach that should be used in evaluating amendments and renewal applications.

5.7.4.3   Acceptance Criteria

The methodologies are acceptable if they meet the following criteria:

(1)   The methodologies proposed to determine the intake of radioactive materials by personnel in work areas where airborne radioactive materials could exist are in accordance with 10 CFR 20.1204 and 20.1201.

(2)   Exposure calculations for natural uranium are consistent with Regulatory Guide 8.30, Section 3 (NRC, 2002).

(3)   For airborne radon daughter exposure (working levels), calculations are consistent with Regulatory Guide 8.30 (NRC, 2002) and Regulatory Guide 8.34, Section C (NRC, 1992a).

(4)   Calculations and guidance for prenatal and fetal radiation exposure are consistent with Regulatory Guide 8.36, "Radiation Dose to the Embryo/Fetus" (NRC, 1992b) and Regulatory Guide 8.13, "Instruction Concerning Prenatal Radiation Exposure" (NRC, 1999).

Operations

(5)   Exposure calculations are presented for routine operations, non-routine operations, maintenance, and cleanup activities and are consistent with Draft Regulatory Guide 8.30 (NRC, 2002) and Regulatory Guide 8.34, Section C (NRC, 1992a).

(6)   Parameters used in exposure calculations are representative of conditions at the site and include the time-weighted exposure that incorporates occupancy time and average airborne concentrations.

        For example, the time of exposure may be arbitrarily set at 40 hours per week; however, workers at some facilities may regularly work longer shifts. Both full-time and part-time employees should be considered in these calculations.

(7)   Estimation of airborne uranium concentrations takes into account the maximum production capacity requested in the application and the anticipated efficiencies of airborne particulate control systems reviewed using in Sections 4.1 and 5.7.1 of this standard review plan.

(8)   All reporting and record keeping of worker doses is done in conformance with Regulatory Guide 8.7 (NRC, 1982) and 10 CFR 20.2103.

(9)   For license renewal applications, the historical results of radiation exposure calculations are included through the most recent reporting period preceding the submittal of the application. The effectiveness of historical radiation exposure calculations is discussed with regard to applicable 10 CFR Part 20 regulatory requirements. Long-term trends are discussed, and any short-term deviations from the long-term trend are explained.

5.7.4.4   Evaluation Findings

If the staff review, as described in this section, results in the acceptance of the exposure calculations, the following conclusions may be presented in the technical evaluation report.

NRC has completed its review of the exposure calculations at the _____ in situ leach facility. This review included an evaluation using the review procedures in standard review plan Section 5.7.4.2 and the acceptance criteria outlined in standard review plan Section 5.7.4.3.

The applicant has provided acceptable techniques for exposure calculations at the _____ in situ leach site. The applicant has techniques to determine intake of radioactive materials by personnel in work areas. The applicant exposure calculations for natural uranium and airborne radon daughter exposure are acceptable and are in conformance with the guidance in Regulatory Guide 8.30 (NRC, 2002) and Regulatory Guide 8.34 (NRC, 1992a). The applicant has acceptable methods to calculate prenatal and fetal radiation exposures consistent with Regulatory Guides 8.13 (NRC, 1999) and 8.36 (NRC, 1992b). All exposure calculation methods for routine operations, non-routine operations, maintenance, and cleanup activities are acceptable and are consistent with Regulatory Guide 8.30 (NRC, 2002) and Regulatory Guide 8.34 (NRC, 1992a). The applicant has used parameters that are representative of the site, such as using both full- and part-time workers in exposure calculations. The applicant has considered maximum production capacity and anticipated efficiencies of airborne particulate control systems in exposure calculations. All reporting and record keeping is in conformance with Regulatory Guide 8.7 (NRC, 1982).

Based on the information provided in the application and the detailed review conducted of the exposure calculations at the _____ in situ leach facility, the staff has concluded that the exposure calculations are acceptable and are in compliance with 10 CFR 20.1101, which defines radiation protection program requirements; 10 CFR 20.1201(a), which specifies individual occupational dose limits; 10 CFR 20.1201(e), which defines allowed intake of soluble uranium; 10 CFR 20.1202, which describes the means of compliance when summing internal and external doses; 10 CFR 20.1203, for determination of dose from airborne external radiation; 10 CFR 20.1204, which provides requirements for determination of internal exposure; and 10 CFR 20.1208, which specifies the exposure limits for a fetus.

5.7.4.5   References

NRC. Regulatory Guide 8.30, "Health Physics Surveys in Uranium Recovery Facilities." Washington, DC: NRC, Office of Nuclear Regulatory Research. 2002.

———. Regulatory Guide 8.13, "Instruction Concerning Prenatal Radiation Exposure." Revision 3. Washington, DC: NRC, Office of Standards Development. 1999.

———. Regulatory Guide 8.34, "Monitoring Criteria and Methods To Calculate Occupational Radiation Doses." Washington, DC: NRC, Office of Standards Development. 1992a.

———. Regulatory Guide 8.36, "Radiation Dose to the Embryo/Fetus." Washington, DC: NRC, Office of Standards Development. 1992b.

———. Regulatory Guide 8.7, "Instructions for Recording and Reporting Occupational Radiation Exposure Data." Revision 1. Washington, DC: NRC, Office of Standards Development. 1982.

5.7.5 Bioassay Program

5.7.5.1   Areas of Review

The staff should review descriptions of the bioassay program and how the bioassay results will be used to confirm results derived from the airborne radiation monitoring program (standard review plan Section 5.7.3) and the exposure calculations (standard review plan Section 5.7.4). The staff should review the criteria for including workers in the bioassay program, the types and frequencies of bioassays performed, and action levels applied to the results.

5.7.5.2   Review Procedures

The staff should determine whether the bioassay program is adequate to confirm results determined in the airborne radiation monitoring program (standard review plan Section 5.7.3) and the exposure calculations (standard review plan Section 5.7.4). The staff should review the

Operations

bioassay program to ensure that it is consistent with applicable sections of Regulatory Guide 8.22, "Bioassay at Uranium Mills" (NRC, 1988). The staff review should check to ensure that all workers who are routinely exposed to yellowcake dust are included in the bioassay program and that sampling and analysis frequencies are sufficient to detect and take corrective action against high intakes of uranium in the workplace. Primarily, the program should involve workers stationed in yellowcake drying areas and those who conduct regular maintenance on drying and ventilation/filtration equipment.

For license renewals and amendment applications, Appendix A to this standard review plan provides guidance for examining facility operations and the approach that should be used in evaluating amendments and renewal applications.

5.7.5.3   Acceptance Criteria

The bioassay program is acceptable if it meets the following criteria:

(1)   It is consistent with applicable sections of Regulatory Guide 8.22 (NRC, 1988) and Regulatory Guide 8.31 (NRC, 2002) including as low as is reasonably achievable requirements. The bioassay program is adequate to confirm results determined from the airborne radiation monitoring program (standard review plan Section 5.7.3) and the exposure calculations (standard review plan Section 5.7.4).

(2)   The determination of which workers will be monitored in the bioassay program is consistent with Regulatory Guide 8.22, Section 2 (NRC, 1988).

(3)   Sampling and analysis frequencies include baseline urinalyses for all new employees and exit bioassays on termination of employment and are consistent with Regulatory Guide 8.22, Section 4 (NRC, 1988) and Regulatory Guide 8.9, Revision 1, "Acceptable Concepts, Equations, and Assumptions for a Bioassay Program" (NRC, 1993).

(4)   Action levels for bioassay monitoring are set in accordance with Regulatory Guide 8.22, Section 5 (NRC, 1988).

(5)   All reporting and record keeping are done in conformance with the requirements of 10 CFR Part 20, Subpart L and Subpart M.

(6)   For license renewal applications, the historical bioassay program results are included through the most recent reporting period preceding the submittal of the application. The effectiveness of the historical program is discussed with regard to all applicable 10 CFR Part 20 regulatory requirements. Long-term trends are discussed, and any short-term deviations from the long-term trend are explained.

5.7.5.4   Evaluation Findings

If the staff review, as described in this section, results in the acceptance of the bioassay program, the following conclusions may be presented in the technical evaluation report.

NRC has completed its review of the bioassay program at the _____ *in situ* leach facility. This review included an evaluation using the review procedures in standard review plan Section 5.7.5.2 and the acceptance criteria outlined in standard review plan Section 5.7.5.3.

The applicant has established an acceptable bioassay program at the _____ *in situ* leach site that is consistent with Regulatory Guide 8.22 (NRC, 1988). An acceptable program for baseline urinalysis and exit bioassay is in place. Individuals routinely exposed to yellowcake dust are a part of the bioassay program. An acceptable action program to curtail uranium intake is established, and appropriate actions levels are set. The applicant has established reporting and record keeping protocols in conformance with the requirements of 10 CFR Part 20, Subpart L.

Based on the information provided in the application and the detailed review conducted of the bioassay program at the _____ *in situ* leach facility, the staff concludes that the bioassay program is acceptable and is in compliance with 10 CFR 20.1204, which provides requirements for the determination of internal exposure; and 10 CFR Part 20, Subpart L, which establishes record keeping requirements.

5.7.5.5   References

NRC. Regulatory Guide 8.31, "Information Relevant to Ensuring that Occupational Radiation Exposures at Uranium Mills Will Be As Low As Reasonably Achievable." Rev. 1. Washington, DC: NRC, Office of Nuclear Regulatory Research. 2002.

————. Regulatory Guide 8.9, "Acceptable Concepts, Models, Equations, and Assumptions for a Bioassay Program." Revision 1. Washington, DC: NRC, Office of Standards Development. 1993.

————. Regulatory Guide 8.22, "Bioassay at Uranium Mills." Revision 1. Washington, DC: NRC, Office of Standards Development. 1988.

5.7.6 Contamination Control Program

5.7.6.1   Areas of Review

The staff should review the contamination control program proposed to prevent employees from entering clean areas or from leaving the site while contaminated with radioactive materials. Levels of radioactive contamination will be monitored by means of a radiation survey program. Review areas include methods for surveying occupational radiation levels, housekeeping and cleanup requirements; specifications in process areas to control contamination; frequency of surveys of clean areas; survey methods; and minimum sensitivity, range, and calibration

Operations

frequency of survey equipment. Proposed contamination criteria or action levels for clean areas and for the release of materials, equipment, and work clothes from clean areas or from the site should be evaluated. The staff should also review the methods proposed to ensure that the licensee reduces residual contamination below limits before authorizing release of equipment for unrestricted use.

5.7.6.2   *Review Procedures*

The staff should determine whether the contamination control program proposed to prevent contaminated employees from entering clean areas or from leaving the site is in conformance with regulatory requirements in Regulatory Guide 8.30 (NRC,2002). Requirements for a contamination control program (e.g., maintaining change areas and personal alpha radiation monitoring before leaving radiation areas) should be included in standard operating procedures and discussed in the application. The staff should confirm that the license applicant has a contamination control program consistent with the guidance on conducting surveys for contamination of skin and personal clothing provided in Regulatory Guide 8.30 (NRC, 2002). The staff should ensure that the licensee eliminates residual contamination on equipment and materials to within acceptable release limits before release of the equipment for unrestricted use.

For license renewals and amendment applications, Appendix A to this standard review plan provides guidance for examining facility operations and the approach that should be used in evaluating amendments and renewal applications.

5.7.6.3   *Acceptance Criteria*

The contamination control program is acceptable if it meets the following criteria:

(1)   Radiation surveys of workers will be conducted to prevent contaminated employees from entering clean areas or from leaving the site in conformance with guidance in Regulatory Guide 8.30 (NRC, 2002).

   The proposed contamination control program is consistent with the guidance on conducting surveys for contamination of skin and personal clothing provided in Regulatory Guide 8.30 (NRC, 2002).

(2)   Requirements for a contamination control program (e.g., maintaining change areas and personal alpha radiation monitoring before leaving radiation areas) are included in standard operating procedures or are discussed in the application.

   These procedures should be consistent with the guidance on conducting surveys for contamination of skin and personal clothing provided in Regulatory Guide 8.30 (NRC, 2002).

(3)   Action levels for surface contamination are set in accordance with Regulatory Guide 8.30, Section 4 (NRC, 2002).

(4)	Monitoring equipment by type, specification of the range, sensitivity, calibration methods and frequency, availability, and planned use is adequately described. The application demonstrates that the ranges of sensitivity for monitoring equipment will be appropriate to expected facility operation.

(5)	All reporting and record keeping is done in conformance with the requirements of 10 CFR Part 20, Subpart L and Subpart M.

(6)	The licensee will ensure that radioactivity on equipment or surfaces is not covered by paint, plating, or other covering material unless contamination levels, as determined by a survey and documented, are below the limits specified in Table 5.7.6.3-1 of this standard review plan before application of the covering. A reasonable effort will be made to minimize the contamination before the use of any covering.

(7)	The radioactivity of the interior surfaces of pipes, drain lines, or duct work will be determined by making measurements at all traps and other appropriate access points, provided that contamination at these locations is likely to be representative of contamination on the interior of the pipes, drain lines, or duct work.

(8)	The licensee will make a comprehensive radiation survey, in conformance with Regulatory Guide 8.30, Section 1 (NRC, 2002) and NUREG–1575, Revision 1 (NRC, 2000) "Multi-Agency Survey and Site Investigation Manual (MARSSIM)" that establishes that contamination is within the limits specified in Table 5.7.6.3-1 and is as low as is reasonably achievable before release of equipment or scrap for unrestricted use.

(9)	Appropriate criteria are established to relinquish possession or control of equipment or scrap having surfaces contaminated with material in excess of the limits specified in Table 5.7.6.3-1:

(a)	The applicant will provide detailed information describing the equipment, or scrap; the radioactive contaminants; and the nature, extent, and degree of residual surface contamination.

(b)	The applicant will provide a detailed health and safety analysis that reflects that the residual amounts of contaminated materials on surface areas, together with other considerations such as prospective use of the equipment, or scrap, are unlikely to result in an unreasonable risk to the health and safety of the public.

(c)	The applicant includes materials created by special circumstances including, but not limited to, the razing of buildings, transfer of structures or equipment, or conversion of facilities to a long-term storage facility or to standby status.

Table 5.7.6.3-1. Acceptable Surface Contamination Levels (U.S. Atomic Energy Commission, 1974)			
Nuclides[a]	Average[b,c,d]	Maximum[b,d,e]	Removable[b,d,f]
Natural Uranium, Uranium-235, -238, and associated decay products	5,000 α dpm/100 cm^2	15,000 α dpm/100 cm^2	1,000 α dpm/100 cm^2
Transuranics, Radium-226, Radium-228, Thorium-230, Thorium-118, Protactinium-231, Actinium-227, Iodine-125, Iodine-129	100 dpm/100 cm^2	300 dpm/100 cm^2	20 dpm/100 cm^2
Natural Thorium, Thorium-232, Strontium-90, Radium-223, -224, Uranium-232, Iodine-126, Iodine-131, Iodine-133	1,000 dpm/100 cm^2	3,000 dpm/100 cm^2	200 dpm/100 cm^2
Beta-gamma emitters (nuclides with decay modes other than alpha emission or spontaneous fission) except Strontium-90, and others noted above	5,000 dpm/100 cm^2	15,000 dpm/100 cm^2	1,000 dpm/100 cm^2

[a] Where surface contamination by both alpha- and beta-gamma-emitting nuclides exists, the limits established for alpha- and beta-gamma-emitting nuclides should apply independently.
[b] As used in this table, dpm (disintegrations per minute) means the rate of emission by radioactive material as determined by correcting the counts per minute observed by an appropriate factor for background, efficiency, and geometric factors associated with the instrumentation.
[c] Measurements of average contamination should not be averaged over more than 1 m^2. For objects of less surface area, the average should be derived for each such object.
[d] The average and maximum radiation levels associated with surface contamination resulting from beta-gamma emitters should not exceed 0.2 mrad/hr at 1 cm and 1.0 mrad/hr at 1 cm, respectively, measured through not more than 7 mg/cm^2 of total absorber.
[e] The maximum contamination level applies to an area of not more than 100 cm^2.
[f] The amount of removable radioactive material per 100 cm^2 of surface area should be determined by wiping that area with dry filter or soft absorbent paper, applying moderate pressure, and assessing the amount of radioactive material on the wipe with an appropriate instrument of known efficiency. When removable contamination on objects of less surface area is determined, the pertinent levels should be reduced proportionally and the entire surface should be wiped.
Reference: U.S. Atomic Energy Commission. Regulatory Guide 1.86, "Termination of Operating Licenses for Nuclear Reactors." Washington, DC: U.S. Atomic Energy Commission. June 1974.

5.7.6.4   Evaluation Findings

If the staff review, as described in this section, results in the acceptance of the contamination control program, the following conclusions may be presented in the technical evaluation report and environmental assessment.

NRC has completed its review of the contamination control program at the _____ in situ leach facility. This review included an evaluation using the review procedures in standard review plan Section 5.7.6.2 and the acceptance criteria outlined in standard review plan Section 5.7.6.3.

The applicant has established an acceptable contamination control program at the _____ in situ leach site. Acceptable controls are in place to prevent contaminated employees from entering clean areas or from leaving the site. The standard operating procedures will include provisions for contamination control, such as maintaining changing areas and personal alpha radiation monitoring before leaving radiation areas. Acceptable action levels have been set in accordance with Regulatory Guide 8.30 (NRC, 2002), and plans for surveys are in place for skin and personal clothing contamination. The applicant has established that all items removed from the restricted area are surveyed by the radiation safety staff and meet release limits. All reporting and record keeping is done in conformance with protocols established in Regulatory Guide 8.7 (NRC, 1982). The applicant has demonstrated that the range, sensitivity, and calibration of monitoring equipment will protect the health and safety of employees during the full scope of facility operations. The licensee has demonstrated that contaminated surfaces will not be covered unless, before covering, a survey documents that the contamination level is below the limits specified in Table 5.7.6.3-1. The applicant will determine the radioactivity on the interior surfaces of pipes, drain lines, or duct work by making measurements at appropriate access points that will have been shown to be representative of the interior contamination. The applicant has committed to establishing that contamination on equipment, or scrap will be within the limits in Table 5.7.6.3-1 before unrestricted release. To relinquish possession or control of equipment, or scrap with material in excess of the limits specified in Table 5.7.6.3-1, the applicant will provide detailed information on the contaminated material, provide a detailed health and safety analysis that shows that the release of the contaminated material will not result in an unreasonable risk to the health and safety of the public, and obtain NRC staff approval.

Based on the information provided in the application and the detailed review conducted of the contamination control program at the _____ in situ leach facility, the staff concludes that the contamination control program is acceptable and is in compliance with 10 CFR 20.1101, which defines radiation protection program and as low as is reasonably achievable requirements; 10 CFR 20.1501, which provides survey and monitoring requirements; and 10 CFR 20.1702, which allows employees to limit dose to individuals by controlling access, limiting exposure times, prescribing use of respiratory equipment, or other controls.

Operations

5.7.6.5   References

NRC. Regulatory Guide 8.30, "Health Physics Surveys in Uranium Recovery Facilities." Washington, DC: NRC, Office of Nuclear Regulatory Research. 2002.

———. NUREG–1575, "Multi-Agency Radiation Survey and Site Investigation Manual (MARSSIM)." Revision 1. Washington, DC: NRC. 2000.

———. Regulatory Guide 8.7, "Instructions for Recording and Reporting Occupational Radiation Exposure Data." Revision 1. Washington, DC: NRC, Office of Standards Development. 1982.

5.7.7   Airborne Effluent and Environmental Monitoring Program

5.7.7.1   Areas of Review

The staff should review the airborne effluent and environmental monitoring programs proposed for measuring concentrations and quantities of both radioactive and non-radioactive materials released to and in the environment surrounding the facility. The staff should review the technical bases proposed for determining environmental concentrations for demonstrating compliance with standards. The staff review should focus on the frequency of sampling and analysis, the types and sensitivity of analysis, action levels and corrective action requirements, the minimum number and criteria for locating effluent and environmental monitoring stations, and the commitments for semiannual effluent and environmental monitoring reporting. The staff should review a topographic map of the site and the surrounding area showing monitoring locations.

5.7.7.2   Review Procedures

The staff should determine whether the proposed environmental monitoring programs are sufficient to limit exposures and releases of radioactive and hazardous materials as required by 10 CFR 20.2007.

The staff should determine whether the airborne effluent and environmental monitoring programs proposed for measuring concentrations and quantities of both radioactive and hazardous materials released to and in the environment around the proposed facility are in accordance with the regulatory requirements.

The staff should ensure that the license applicant has adequately considered site-specific aspects of climate and topography in determining locations for off-site airborne effluent monitoring stations and environmental sampling areas such that they are capable of detecting maximum concentrations expected from facility operations in the environment. In conducting its review, the staff should refer to guidance in Regulatory Guide 4.14, Revision 1 (NRC, 1980) which contains information on determining sampling locations, types, methods, frequencies, and analyses that are sufficient to comply with the applicable requirements for protection of the public from off-site exposures.

The reviewer shall confirm that the applicant has committed to adequate semiannual airborne effluent and environmental monitoring reporting.

For license renewals and amendment applications, Appendix A to this standard review plan provides guidance for examining facility operations and the approach that should be used in evaluating amendments and renewal applications.

5.7.7.3 Acceptance Criteria

The airborne effluent and environmental monitoring program is acceptable if it meets the following criteria:

(1)   The proposed airborne effluent and environmental monitoring program is consistent with Regulatory Guide 4.14, Sections 1.1 and 2.1 (NRC, 1980) and as low as is reasonably achievable requirements as described in Regulatory Guide 8.37, Section 3 (NRC, 1993).

(2)   The proposed locations of the airborne effluent monitoring stations are consistent with guidance in Regulatory Guide 4.14, Sections 1.1.1 and 2.1.2 (NRC, 1980).

The license applicant adequately considers site-specific aspects of climate and topography in determining the number and locations of off-site airborne monitoring stations and environmental sampling areas. The criteria used in selecting sampling locations should be given. All sampling locations should be clearly shown relative to the proposed facility, nearest residences, and population centers on topographic maps of the appropriate scale.

(3)   The proposed airborne effluent and environmental monitoring program should sample radon, air particulates, surface soils, subsurface soils, vegetation, direct radiation, and sediment in accordance with Regulatory Guide 4.14, Section 3 (NRC, 1980).

(4)   The proposed sampling methods are consistent with guidance in Regulatory Guide 4.14, Section 3 (NRC, 1980).

(5)   For license renewal applications, the historical airborne effluent and environmental monitoring program results are included through the most recent reporting period preceding the submittal of the application. The effectiveness of the historical program is discussed with regard to all applicable regulatory requirements. Long-term trends are discussed, and any short-term deviations from the long-term trend are explained.

(6)   The applicant commits to semiannual airborne effluent and environmental monitoring reporting. These reports will be submitted to the appropriate NRC Regional Office with copies to the Chief, Fuel Cycle Facilities Branch and the project manager. The reports will specify the quantity of each of the principal radionuclides released to unrestricted areas in liquid and gaseous effluents during the previous 6 months, injection rates, recovery rates, injection manifold pressures, and injection trunk line pressures for each

Operations

        satellite facility. The process rate and pressure data are to be reported as monthly averages. A license condition will be imposed to specify these reporting requirements.

5.7.7.4   Evaluation Findings

If the staff review, as described in this section, results in the acceptance of the airborne effluent and environmental monitoring programs, the following conclusions may be presented in the technical evaluation report and environmental assessment.

NRC has completed its review of the airborne effluent and environmental monitoring programs at the _____ in situ leach facility. This review included an evaluation using the review procedures in standard review plan Section 5.7.7.2 and the acceptance criteria outlined in standard review plan Section 5.7.7.3.

The applicant has established acceptable airborne effluent and environmental monitoring programs at the _____ in situ leach site. The programs are consistent with guidance in Regulatory Guide 4.14 (NRC, 1980). The applicant will sample radon, air particulates, surface soils, subsurface soils, vegetation, direct radiation, and sediment. Locations of monitoring stations are consistent with Regulatory Guide 4.14 (NRC, 1980). Instrumentation is appropriate.

Based on the information provided in the application and the detailed review conducted of the airborne effluent and environmental monitoring programs at the _____ in situ leach facility, the staff concludes that the airborne effluent and environmental monitoring programs are acceptable and are in compliance with 10 CFR 20.1302, which requires effluent monitoring to determine dose to individual members of the public; 10 CFR 20.1501, which specifies survey and monitoring requirements; 10 CFR Part 20, Subpart L, which establishes record keeping requirements; and10 CFR 40.65, which specifies effluent and environmental monitoring requirements.

5.7.7.5   References

NRC. Regulatory Guide 8.37, "ALARA Levels for Effluent from Materials Facilities." Washington, DC: NRC, Office of Standards Development. 1993.

———. Regulatory Guide 4.14, "Radiological Effluent and Environmental Monitoring at Uranium Mills." Revision 1. Washington, DC: NRC, Office of Standards Development. 1980.

5.7.8   Ground-Water and Surface-Water Monitoring Programs

5.7.8.1   Areas of Review

There are three distinct phases of ground-water and surface-water monitoring: pre-operational, operational, and restoration. Pre-operational monitoring is conducted as a part of site characterization, and review procedures are in Section 2 of this standard review plan. Restoration monitoring is conducted during the ground-water restoration phase of operations, and review procedures are in Section 6. This standard review plan section deals specifically with monitoring ground-water and surface-water quality during the production or operational phase of *in situ* leach activities.

The staff should review the technical bases and procedures for the following components of an effective ground-water and surface-water operational monitoring program:

(1)   Well field baseline water quality monitoring programs (ground-water and surface-water)

(2)   Selection of excursion indicators and their respective upper control limits

(3)   The placement of excursion monitoring wells

(4)   Well field testing to verify horizontal continuity between the production zone and perimeter wells and vertical isolation between the production zone and vertical excursion monitor wells

(5)   The excursion monitoring program, including well sampling schedules, criteria for placing well fields on excursion status, and corrective actions to be taken in the event of an excursion

(6)   The surface-water monitoring program

For all of the preceding aspects of ground-water and surface-water monitoring programs that involve analysis of water samples, procedures for sample collection and analysis should be reviewed.

5.7.8.2   *Review Procedures*

Well field hydrologic and water chemistry data are collected before *in situ* leach operations to establish a basis for comparing operational monitoring data. Hydrologic data are used to (i) evaluate whether the well field can be operated safely, (ii) confirm monitor wells have been located correctly, and (iii) design aquifer restoration activities. Water chemistry data are used to establish a set of water quality indicators, and the concentrations of these indicators in monitoring wells are used to determine whether the well field is being operated safely. Water chemistry data are also used to set the water quality standard for restoring the production zones and adjacent aquifers after *in situ* leach extraction ceases. The reviewer should determine whether these objectives of the operational monitoring program have been met. To this end, the reviewer should

(1)   Verify that procedures for establishing baseline water quality include acceptable sample collection methods, a set of sampled parameters that is appropriate for the site and *in situ* leach extraction method, and collection of sample sets that are sufficient to represent any natural spatial and temporal variations in water quality.

(2)   Review the applicant's selection (or procedure for selecting) the set of water quality parameters and their respective upper control limits that will be used as indicators to

Operations

> ensure timely detection and reporting of unplanned lixiviant migration (excursions) from production zones.

(3)   Review the applicant's technical basis or procedures for establishing the appropriate monitor well spacing for vertical and horizontal excursion monitoring.

(4)   Evaluate whether well field testing is sufficient to show a horizontal hydraulic connection between the production zones and the perimeter monitor well network, and vertical hydraulic separation between the production zones and the shallow and deep monitor wells.

(5)   Evaluate whether procedures describing the operational excursion monitoring program include sampling schedules, sampling and analytical procedures, criteria for placing well fields on excursion status, and corrective action and notification procedures to be followed if an excursion is detected.

(6)   Evaluate whether a surface-water monitoring program is necessary at the site and, if so, whether the monitoring program will be effective to detect migration of contaminants into surface-water bodies.

In conducting these evaluations, the reviewer should consider the review of ground-water activities conducted by state and other federal agencies to identify any areas where dual reviews can be eliminated. Although the staff must make the necessary findings of compliance with applicable regulations, if a state or other federal agency asks questions in a particular area, the reviewer need not duplicate those questions. Instead, the reviewer can rely on the answers to the state or federal agency questions if they are acceptable, and if the applicant submits them as part of the NRC application. The reviewer should make every effort to coordinate the NRC technical review with the state or other federal agency with overlapping authority to avoid unnecessary duplication of effort.

For license renewals and amendment applications, Appendix A to this standard review plan provides guidance for examining facility operations and the approach that should be used in evaluating amendments and renewal applications.

5.7.8.3   *Acceptance Criteria*

The ground-water and surface-water monitoring program should ensure that an excursion is detected long before *in situ* leach solutions could seriously degrade the quality of ground-water outside the well field area. Early detection of excursions by a monitor well is influenced by the thickness of the aquifer monitored, the distance that monitor wells are placed from the well field and from each other, the frequency that the monitor wells are sampled, the water quality parameters that are sampled, and the concentrations of parameters that will be used to declare that an excursion has been detected.

The ground-water and surface-water monitoring programs are acceptable if they will allow the early detection and timely restoration of excursions. The following criteria must be met by *in situ* leach operational monitoring programs:

(1)   For each new well field, the applicant's approach for establishing baseline water quality data is sufficient to (i) define the primary restoration goal of returning each well field to its pre-operational water quality conditions and (ii) provide a standard for determining when an excursion has occurred. The reviewer should verify that acceptable procedures were used to collect water samples, such as American Society for Testing and Materials D4448 (American Society for Testing and Materials, 1992). The reviewer should also ensure that acceptable statistical methods are used to meet these three objectives, such as American Society for Testing and Materials D6312 (American Society for Testing and Materials, 1998).

Baseline sampling programs should provide enough data to adequately evaluate natural spatial and temporal variations in pre-operational water quality. At least four independent sets of samples should be collected, with adequate time between sets to represent any pre-operational temporal variations. A set of samples is defined as a group of at least one sample at each of the designated baseline monitor wells and analyzed for the water quality conditions of the sampled aquifer at a specific time.

An acceptable set of samples should include all well field perimeter monitor wells, all upper and lower aquifer monitor wells, and at least one production/injection well per acre in each well field. For large well fields, it may not be practical to sample one production/injection well per acre. Consequently, enough production/injection wells must be sampled to provide an adequate statistical population if fewer than one well per acre is used. As a general guideline, for normally and log-normally distributed populations, at least six samples are required to achieve 90 percent confidence that any random sample will lie within two standard deviations from the sample mean. In no case should the baseline sampling density for production/injection wells be less than one per 4 acres.

The applicant should identify the list of constituents sampled for baseline concentrations. Table 2.7.3-1 provides a list of acceptable constituents for monitoring at *in situ* leach facilities. Alternatively, applicants may propose a list of constituents that is tailored to a particular location. In such cases, sufficient technical bases must be provided to demonstrate the acceptability of the selected constituent list. For example, many licensees have decided not to sample for Th-230; Th-230 is a daughter product from the decay of uranium-238, and studies have shown that it is mobilized by bicarbonate-laden leaching solutions. However, studies have also shown that after restoration, thorium in the ground-water will not remain in solution, because the chemistry of thorium causes it to precipitate and chemically react with the rock matrix (Hem, 1985). As a result of its low solubility in natural waters, thorium is found in only trace concentrations. Additionally, chemical tests for thorium are expensive, and are not commonly included in water analyses at *in situ* leach facilities. This example concerning Th-230 demonstrates an acceptable technical basis for excluding Th-230 from the list of sampled constituents. For all constituents that are sampled, laboratory reports documenting the measurements should be maintained by the applicant.

An outlier is a single non-repeating value that lies far above or below the rest of the sample values for a single well. Dealing with outliers in the sample sets should be done

Operations

using proper statistical methods. The outlier may represent a sampling, analytical, or other unknown source of error or an unidentified randomness in the data. Its inclusion within the sample could significantly change the baseline data, since the outlier is not typical of the bulk of the samples. All calculations, assumptions, and conclusions made by the applicant in evaluating outliers should be fully explained. When an outlier is suspected, perhaps the easiest solution is to take another sample from the source well; if the repeat sample yields the same results, then the outlier should not be discarded. If the repeat sample is more consistent with the statistical population, the outlier can be replaced with the new sample. Another acceptable method for dealing with potential outliers is to accept any value within three standard deviations of the mean (the standard deviation should be calculated without using the suspected outliers). It is often necessary to perform log transformations on data to better approximate a normal distribution before calculating sample statistics. Care should be taken not to exclude suspected outliers that ultimately may represent bimodal distributions. Methods in American Society for Testing and Materials E178 (American Society for Testing and Materials, 1994), NUREG/CR–4604 (NRC, 1988) and NUREG–1475 (NRC, 1994) are acceptable methods for outlier calculation. Other documented and technically justified methods used by applicants will be considered in the evaluation of outliers (e.g., EPA, 1989).

(2)   The applicant selects excursion indicator constituents and upper control limits. Upper control limits are concentrations for excursion indicator constituents that provide early warning that leaching solutions are moving away from the well fields and that ground-water outside the monitor well ring may be threatened. Excursion indicator constituents should be parameters that are strong indicators of the *in situ* leach process and that are not significantly attenuated by geochemical reactions in the aquifers. If possible, the chosen parameters should be easily analyzed to allow timely data reporting. The upper control limit concentrations of the chosen excursion indicators should be set high enough that false positives (false alarms from natural fluctuations in water chemistry) are not a frequent problem, but not so high that significant ground-water quality degradation could occur by the time an excursion is identified. A minimum of three excursion indicators should be proposed. The choice of excursion indicators is based on lixiviant content and ground-water geochemistry. Ideal excursion indicators are measurable parameters that are found in significantly higher concentrations during *in situ* leach operations than in the natural waters. At most uranium *in situ* leach operations, chloride is an excellent excursion indicator because it acts as a conservative tracer, it is easily measured, and chloride concentrations are significantly increased during *in situ* leaching. Conductivity, which is correlated to total dissolved solids, is also considered to be a good excursion indicator (Staub, 1986; Deutsch, 1985). Total alkalinity (carbonate plus bicarbonate plus hydroxide) is an excellent indicator in well fields where sodium bicarbonate or carbon dioxide is used in the lixiviant. If conductivity is used to estimate total dissolved solids, measurements will be normalized to a reference temperature, usually 25 °C, because of the temperature dependence of conductivity.

Calcium, sodium, and sulfate are usually found at significantly higher levels in *in situ* solutions than in natural ground-water concentrations. The use of cations

5-40

(e.g., calcium^{2+}, sodium$^+$) as excursion indicators is generally not appropriate because they are subject to ion exchange with the host rock. The use of sulfate may give false alarms because of induced oxidation around a monitor well (Staub, 1986; Deutsch, 1985). However, this should only be a problem if upper control limit values are set too conservatively. Uranium is not considered a good excursion indicator because, although it is mobilized by *in situ* leaching, it may be retarded by reducing conditions in the aquifer. Although water level changes in artesian aquifers are quickly transmitted, water levels are generally not considered good indicators, because water levels tend to have significant natural variability. The applicant may choose to add a non-reactive, conservative tracer to *in situ* leach solutions to act as an excursion indicator. The applicant is required to provide the technical bases for the selection of excursion indicators.

Upper control limit concentrations must be set to easily identify excursions. An excursion is defined to occur whenever two or more excursion indicators in a monitoring well exceed their upper control limits. The upper control limit for each excursion indicator must generally be less than the lowest concentration that typically occurs in the lixiviant while the well field is in operation. Each upper control limit must also be greater than the baseline concentration for its respective excursion indicator. Applicant site-specific experience is often valuable in determining appropriate upper control limits that provide timely detection and avoid false alarms. Guidance for appropriate statistical methods that can be used to establish upper control limits can be found in American Society for Testing and Materials D6312 (American Society for Testing and Materials, 1998).

Upper control limits for a specific excursion indicator should be determined on a statistical basis to account for likely spatial and temporal concentration variations within the mineralized zone. Statistical techniques, such as the student's t-test, are acceptable for setting upper control limits. In some cases, the use of a simple percentage increase above baseline values is acceptable. The staff has decided that in areas with good water quality (a total dissolved solids less than 500 mg/L), setting the upper control limit at a value of 5 standard deviations above the mean of the measured concentrations is an acceptable approach. However, in some aquifers of good water quality, low chloride concentrations have been found to have such a narrow statistical distribution that a specified concentration (e.g., 15 mg/L) above the mean or the mean plus 5 standard deviations approach, which ever is greater, has been used to establish the chloride upper control limit.

The same upper control limits may be assigned to all monitor wells within a particular hydrogeologic unit in a given well field if baseline data indicate little chemical heterogeneity. Alternatively, if individual monitor wells in a given unit exhibit unique baseline water quality, upper control limits may be assigned on a well-by-well basis. If upper control limits vary from well to well, a table should be included listing all monitor wells and their respective upper control limits.

(3)   The applicant establishes criteria for determining monitor well locations. Production zone perimeter monitor wells are used to detect horizontal excursions outside the well

field boundary. They generally surround the entire well field and are screened over the entire production zone hydrogeologic unit. Perimeter monitor wells should be placed close enough to the well field to provide timely detection, yet they should be far enough away from the well field to avoid numerous false alarms. Previously approved *in situ* leach excursion monitoring systems used monitor wells as far as 180 m [600 ft] and as near as 75 m [250 ft] from the well field edge (NRC, 2001, Table 4-6). The licensee should be afforded some discretion in determining the appropriate distance of horizontal excursion monitor wells from the well field, but should provide justification for distances greater than about 150 m [500 ft]. For example, a rigorous modeling demonstration that a theoretical excursion can be controlled at the monitor well locations within 60 days of detection is an acceptable technical basis. The horizontal excursion monitor wells must be spaced close enough to one another so that the likelihood of missing an excursion plume is low. In determining the appropriate spacing between perimeter monitoring wells, the applicant must consider such factors as the distance of the monitoring wells from the edge of the well field, the minimum likely size of an excursion source zone, ground-water flow directions and velocities outside of the well field, and the potential for mixing and dispersion. Staff should consult NUREG/CR–6733 (NRC, 2001, Section 4.3.3) for an analysis and discussion of acceptable approaches for establishing the appropriate monitor well spacing.

NUREG/CR–6733 (NRC, 2001, Section 4.3.3) established that significant risks for vertical excursions may exist if monitor wells are randomly located, given the typical criteria for spacing of vertical excursion monitor wells at licensed *in situ* leach facilities {e.g., one well per 1.6 ha [4 acres] for overlying aquifers; one well per 3.2 ha [8 acres] for underlying aquifers}. Thus, location of vertical excursion monitor wells within the well field should be such that the likelihood of detecting a vertical excursion is maximized. The appropriate number of these monitor wells may vary from site to site. It may be appropriate to exclude the requirement to monitor water quality in the underlying aquifer if (i) the underlying aquifer is a poor producer of water, (ii) the underlying aquifer is of poor water quality, (iii) there is a large aquitard between the production zone and the underlying aquifer and few boreholes have penetrated the aquitard, or (iv) deep monitor wells would significantly increase the risk of a vertical excursion into the underlying aquifer. Monitor wells completed in aquifers above the first overlying aquifer may not be required when (i) the aquifers are separated from the production zone by thick aquitards, (ii) a high quality mechanical integrity well testing program will be implemented, or (iii) the aquifers are unsubstantial producers of water or of poor water quality. In well fields where the production zone confining layers are particularly thin, or of questionable continuity, a greater number of monitor wells is appropriate. In general, when the direction of ground-water flow in an upper or lower aquifer is well known, the applicant should consider locating these wells on the hydraulically down gradient side of a well field, in areas where production zone confining layers may be thin or incompetent, and in areas where injection pressure may be highest (i.e., closer to injection wells than to production wells).

The process for determining the screened interval of the monitor wells should be described. Fully screened monitor wells sample the entire thickness of the aquifer. Therefore, excursions could not pass above or below the well screens. However, the concentration of the indicator parameters might be diluted and therefore may not provide timely warning that an excursion is occurring. Partially screened monitor wells only sample the zone of extraction within an aquifer. These wells might miss some excursions, but would suffer less from dilution effects than fully screened wells. For most situations the staff favors fully screened monitor wells. Fully screened monitor wells would assure that excursions will eventually be detected, have the advantage of more accurately representing the water quality that a ground-water user is likely to experience, and do not suffer from the uncertainty of predicting the completion intervals of injection and production wells that have not yet been drilled.

(4)   The applicant establishes well field test procedures. Once a well field is installed, it should be tested to establish that the production and injection wells are hydraulically connected to the perimeter horizontal excursion monitor wells and are hydraulically isolated from the vertical excursion monitor wells. Such testing will serve to confirm the performance of the monitoring system and will verify the validity of the site conceptual model reviewed in Section 2 of this standard review plan. The reviewer should verify that well field test approaches have sound technical bases. Test approaches typically consist of a pumping test that subjects the well field to a sustained maximum withdrawal rate while monitoring the perimeter and vertical excursion wells for drawdown. The test should continue until the effects of pumping can be clearly seen via drawdown in the perimeter monitor wells. Typically, about 0.3 m [1 ft] of drawdown in the perimeter monitor wells will verify hydraulic connection, but the amount may vary because of the distance from the pumping wells, pumping rates, and hydraulic conductivity. To investigate vertical confinement or hydraulic isolation between the production zone and upper and lower aquifers, water levels in upper or lower aquifers may also be monitored during the pumping tests.

(5)   The applicant defines operational approaches for the monitoring program. The monitoring program must indicate which wells will be monitored for excursion indicators, the monitoring frequency, and the criteria for determining when an excursion has occurred. An acceptable excursion monitoring program should indicate that all monitor wells will be sampled for excursion indicators at least every 2 weeks during *in situ* leach operations.

An excursion is deemed to have occurred if two or more excursion indicators in any monitor well exceed their upper control limits. A verification sample must be taken within 48 hours after results of the first analyses were received. If the second sample does not indicate that upper control limits were exceeded, a third sample must be taken within 48 hours after the second set of sampling data was acquired. If neither the second nor the third sample indicates that upper control limits are exceeded, the first sample is considered in error, and the well is removed from excursion status. If either the second or third sample contains indicators above upper control limits, an excursion is confirmed, the well is placed in excursion status, and corrective action must be initiated.

Generally, the risk of contamination to surface-water bodies from *in situ* leach operations is low when proper operational procedures are followed. Any surface-water

5-43

body that lies within the proposed license boundary should be sampled at upstream and downstream locations, both before and during operations. The reviewer should ensure that pre-operational water quality sampling locations for applicable surface-waters are indicated in the application. The pre-operational data should be collected on a seasonal basis for a minimum of 1 year before *in situ* leach operations. Procedures for monitoring surface-water quality during operations should be discussed in the application: this discussion must include a monitoring schedule, monitor locations, and a list of sampled constituents. The applicant may be exempted from monitoring during operations if the site characterization demonstrates that no significant flow of ground water to surface water occurs near the site (e.g., if surface-water bodies are perched and ephemeral).

The excursion monitoring operational procedures must also include corrective action and notification plans in the event of an excursion. NRC must be notified within 24 hours by telephone and within 7 days in writing from the time an excursion is verified. A written report describing the excursion event, corrective actions, and the corrective action results must be submitted to NRC within 60 days of the excursion confirmation. If wells are still on excursion status when the report is submitted, the report must also contain a schedule for submittal of future reports describing the excursion event, corrective actions taken, and results obtained. In the case of a vertical excursion, the report must contain a projected date when characterization of the extent of the vertical excursion would be completed.

Corrective action to retrieve horizontal excursions within the production-zone aquifer is generally accomplished by adjusting the flow rates of the pumping/injection wells to increase process bleed in the area of the excursion. Vertical excursions have proven more difficult to retrieve: at some *in situ* leach facilities, vertical excursions have persisted for years. If an excursion is not corrected within 60 days of confirmation, applicants must either terminate injection of lixiviant into the well field until the excursion is retrieved, or provide an increase to the reclamation surety in an amount that is agreeable to NRC and that would cover the expected full cost of correcting and cleaning up the excursion. The surety increase must remain in force until the excursion is corrected. The written 60-day excursion report should state and justify which course of action will be followed.

If wells are still on excursion status at the time the 60-day report is submitted to NRC, and the surety option is chosen, the well field restoration surety will be adjusted upward. To calculate the increase in surety for horizontal excursions, it is assumed that the entire thickness of the aquifer between the well field and the monitor wells on excursion has been contaminated with lixiviant. The width of the excursion is assumed to be the distance between the monitor wells on excursion status plus one monitor well spacing distance on either side of the excursion. When the excursion is corrected, the additional surety requirements resulting from the excursion will be removed.

To calculate the increase in surety for vertical excursions, an initial estimate of the area contaminated is made. All estimates assume that the entire thickness of the aquifer is contaminated. As characterization of the extent of contamination proceeds, the surety may be increased or decreased, as appropriate. Once the extent of contamination is determined, the area contaminated above background is used to calculate the level of surety. When the vertical excursion is cleaned up, the additional surety requirements resulting from the excursion are removed.

In calculating the increase in surety bonding for horizontal and vertical excursions, the same formula used to calculate the number of pore volumes required to restore a well field is applied to the assumed areas of contamination. This approach is consistent with 10 CFR Part 40, Appendix A, Criterion 9. Increased surety provides assurance that cleanup will be accomplished in the event of licensee default, and surety can be adjusted downward once cleanup is complete. In calculating the area affected by an excursion and the volume of water required to effect restoration, a conservative estimate is taken to ensure that adequate funds are available to clean up the ground water should the licensee fail to do so.

Corrective action for vertical and horizontal excursions can be determined complete when all excursion indicators are below their respective upper control limits, or if only one excursion indicator exceeds its respective upper control limit by less than 20 percent. Stability in the excursion indicator concentrations must be demonstrated by measurements over a suitable time period before the corrective action measures can be discontinued.

(6)   If an *in situ* leach facility is located adjacent to bodies of surface-water, the applicant must establish a surface-water monitoring program that will be effective to detect migration of contaminants into surface-water bodies . Alternatively, the applicant may demonstrate that the risk of contamination from *in situ* leach activities is negligible or that potential releases are within limits set by the Safe Drinking Water Act.

5.7.8.4   *Evaluation Findings*

If the staff review, as described in this section, results in the acceptance of the ground-water and surface-water monitoring programs, the following conclusions may be presented in the technical evaluation report and environmental assessment.

NRC has completed its review of the ground-water and surface-water monitoring programs at the _____ *in situ* leach facility. This review included an evaluation using the review procedures in standard review plan Section 5.7.8.2 and the acceptance criteria outlined in standard review plan Section 5.7.8.3.

The applicant has established acceptable ground-water and surface-water monitoring programs at the _____ *in situ* leach site. The applicant has established acceptable well field baseline sampling programs including the number and timing of samples, constituents sampled, and appropriate statistical methods to remove outliers. The applicant has selected acceptable excursion indicator constituents and an approach for establishing upper control limits. Appropriate criteria are used to establish monitor well locations for all aquifers likely to be affected. Appropriate well field test procedures are established. The applicant has defined acceptable operational approaches for the ground-water and surface-water monitoring

Operations

programs, including identifying appropriate wells for monitoring for excursion indicators, monitoring frequency, and criteria for determining the presence of an excursion. The applicant has defined an acceptable sampling program for any surface-water body that lies within the facility boundary, including downstream sampling locations and standard approaches for monitoring with a schedule and a list of analyzed constituents. The applicant has prepared an acceptable ground-water and surface-water corrective action plan, including notification of NRC and subsequent reporting in the event of an excursion.

Based on the information provided in the application and the detailed review conducted of the ground-water and surface-water monitoring programs at the _____ in situ leach facility, the staff concludes that the ground-water and surface-water monitoring programs are acceptable and are in compliance with 10 CFR 40.32(c), which requires the applicant's proposed equipment, facilities, and procedures to be adequate to protect health and minimize danger to life or property; 10 CFR 40.32(d), which requires that the issuance of the license will not be inimical to the common defense and security or to the health and safety of the public; 10 CFR 40.41(c), which requires the applicant to confine source or byproduct material to the locations and purposes authorized in the license; and 10 CFR 40.31, which defines requirements for applications for specific licenses. The ground-water and surface-water monitoring programs are also in compliance with 10 CFR Part 40, Appendix A, Criteria 5B(1), 5B(5), and 5C, which provide concentration limits for contaminants; 10 CFR Part 40, Appendix A, Criterion 5D, which requires a ground-water corrective action program; and 10 CFR Part 40, Appendix A, Criteria 7 and 7A, which require ground-water monitoring programs.

Pre-operational monitoring is conducted as part of site characterization and is addressed in Section 2 of this technical evaluation report, whereas restoration monitoring is conducted during ground-water restoration and is addressed in Section 6 of this technical evaluation report.

5.7.8.5   References

American Society for Testing and Materials. "Standard Guide for Developing Appropriate Statistical Approaches for Ground-Water Detection Monitoring." Designation D6312–98. West Conshohocken, Pennsylvania: American Society for Testing and Materials. 1998.

––––––. "Standard Practice for Dealing with Outlying Observations." Designation E178. West Conshohocken, Pennsylvania: American Society for Testing and Materials. 1994.

––––––. "Standard Guide for Sampling Groundwater Monitoring Wells." Designation D4448–85a. West Conshohocken, Pennsylvania: American Society for Testing and Materials. 1992.

Deutsch, W.J., et al. NUREG/CR–3709, "Method of Minimizing Ground-Water Contamination From *In Situ* Leach Uranium Mining." Washington, DC: NRC. 1985.

EPA. "Statistical Analysis of Ground-Water Monitoring Data at RCRA (Resource Conservation and Recovery Act) Facilities, Interim Final Guidance." EPA/530–SW–89–026. Washington, DC: EPA. 1989.

Hem, J.D. "Study and Interpretation of the Chemical Characteristics of Natural Water." USGS Water Supply Paper 2254. Third edition. Reston, Virginia: U.S. Geological Survey. 1985.

NRC. NUREG/CR–6733, "A Baseline Risk-Informed, Performance-Based Approach for *In Situ* Leach Uranium Extraction Licensees." Washington, DC: NRC. 2001.

———. NUREG–1475, "Applying Statistics." Washington, DC: NRC. 1994.

———. NUREG/CR–4604, "Statistical Methods for Nuclear Material Management." Washington, DC: NRC. 1988.

Staub, W.P., et al. NUREG/CR–3967, "An Analysis of Excursions at Selected *In Situ* Uranium Mines in Wyoming and Texas." Washington, DC: NRC. 1986.

5.7.9 Quality Assurance

5.7.9.1 Areas of Review

The staff should review the quality assurance programs proposed for all radiological, effluent, and environmental (including ground water) monitoring programs.

5.7.9.2 Review Procedures

The staff should determine whether the quality assurance program proposed by the applicant is sufficient to limit radiation exposures and radioactive releases to as low as is reasonably achievable and is in conformance with regulatory requirements identified in 10 CFR Part 20. The staff should determine if the quality assurance programs proposed for all radiological, effluent, and environmental (including ground water) monitoring are in accordance with Regulatory Guide 4.14, "Radiological Effluent and Environmental Monitoring at Uranium Mills, Revision 1 (NRC, 1980) and Regulatory Guide 4.15, "Quality Assurance for Radiological Monitoring Programs (Normal Operations)—Effluent Streams and the Environment, Revision 1" (NRC, 1979).

For license renewals and amendment applications, Appendix A to this standard review plan provides guidance for examining facility operations and the approach that should be used in evaluating amendments and renewal applications.

5.7.9.3 Acceptance Criteria

The quality assurance program is acceptable if it meets the following criteria:

(1)   The quality assurance program has been established and applied to all radiological, effluent, and environmental programs. The proposed quality assurance plan should be consistent with guidance provided in Regulatory Guide 4.14, Section 3 and 6 (NRC, 1980) and Regulatory Guide 4.15 (NRC, 1979).

Operations

(2)   All reporting and record keeping will be done in conformance with the criteria presented in Section 5.3.2 of this standard review plan.

Note that under the existing 10 CFR Part 20 requirements, a licensee must retain survey and calibration records for 3 years instead of the 2 years mentioned in Regulatory Guide 4.15 (NRC, 1979). Furthermore, existing 10 CFR Part 20 requirements have been updated to include a requirement that all licensees maintain records used to demonstrate compliance and evaluate dose, intake, and releases to the environment until NRC terminates the license.

(3)   For license renewal applications, the historical quality assurance program results are included through the most recent reporting period preceding the submittal of the application. The effectiveness of the historical program is discussed with regard to all applicable 10 CFR Part 20 regulatory requirements. Long-term trends are discussed, and any short-term deviations from the long-term trend are explained.

5.7.9.4   Evaluation Findings

If the staff review, as described in this section, results in the acceptance of the quality assurance program, the following conclusions may be presented in the technical evaluation report and environmental assessment.

NRC has completed its review of the quality assurance program at the _____ *in situ* leach facility. This review included an evaluation using the review procedures in standard review plan Section 5.7.9.2 and the acceptance criteria outlined in standard review plan Section 5.7.9.3.

The applicant has established an acceptable quality assurance program at the _____ *in situ* leach site. The quality assurance program has been applied to all radiological, effluent, and environmental programs consistent with Regulatory Guides 4.14 (NRC, 1980) and 4.15 (NRC, 1979). The applicant has agreed to retain survey and instrument calibration records for 3 years and to retain records to demonstrate compliance and evaluate dose, intake, and releases to the environment until NRC terminates the license.

Based on the information provided in the application and the detailed review conducted of the quality assurance program at the _____ *in situ* leach facility, NRC staff concludes that the quality assurance program is acceptable and is in compliance with 10 CFR 20.1101, which provides requirements for radiation protection programs; 10 CFR Part 20, Subpart L, which specifies record keeping requirements; and 10 CFR Part 20, Subpart M, which defines reporting and notification requirements.

5.7.9.5   References

NRC. Regulatory Guide 4.14, "Radiological Effluent and Environmental Monitoring at Uranium Mills." Revision 1. Washington, DC: NRC, Office of Standards Development. 1980.

——. Regulatory Guide 4.15, "Quality Assurance for Radiological Monitoring Programs (Normal Operations)—Effluent Streams and the Environment." Revision 1. Washington, DC: NRC, Office of Standards Development. 1979.

6.0 GROUND-WATER QUALITY RESTORATION, SURFACE RECLAMATION, AND FACILITY DECOMMISSIONING

6.1 Plans and Schedules for Ground-Water Quality Restoration

It is important to note that the acceptance criteria laid out in this standard review plan are for the guidance of NRC staff responsible for the review of applications to operate *in situ* leach facilities. Review plans are not substitutes for the Commission's regulations, and compliance with a particular standard review plan is not required. This standard review plan provides descriptions of methodologies that have been found acceptable for demonstrating regulatory compliance. Alternative methods and solutions different from those set out in the standard review plan will be acceptable if they provide a basis for the findings requisite to the issuance or continuance of a license by NRC.

In conducting these evaluations, the reviewer should consider the technical evaluations conducted by a state or another federal agency with authorities overlapping those of the NRC. The desired outcome is to identify any areas where duplicative NRC reviews may be reduced or eliminated. The NRC staff must make the necessary evaluations of compliance with applicable regulations for licensing the facility. However, the reviewer may, as appropriate, rely on the applicant's responses to inquiries made by a state or another federal agency to support the NRC evaluation of compliance. The reviewer should make every effort to coordinate the NRC technical review with the state or other federal agency with overlapping authority to avoid unnecessary duplication of effort.

Some of the review methods and acceptance criteria in the following sections are more rigorous than those previously used by the NRC staff. They provide increased confidence in the adequacy of ground-water restoration plans and the sureties associated with them.

Technical assessment of the selected ground-water restoration methods, restoration time and pore volume displacements, and sureties may entail use of detailed, small-scale process models to large-scale, simplified models. Small-scale process models are generally used to evaluate potentially important complexities and mechanisms that govern the evolution of the contaminated areas, while large-scale, simplified models generally consider fewer complexities but may be suitable for evaluating average or effective processes for large areas. Model adequacy should be evaluated regardless of the level of complexity.

This review should be coordinated with the site hydrologic characteristics review conducted using Section 2.7 of this standard review plan.

For license renewals and amendment applications, Appendix A to this standard review plan provides guidance for examining facility operations and the approach that should be used in evaluating amendments and renewal applications.

Ground-Water Quality Restoration, Surface
Reclamation, and Plant Decommissioning

6.1.1 Areas of Review

The staff should review the following aspects of the ground-water quality restoration program:

(1) Ground-water modeling used to estimate restoration time and the extent of uncertainties in processes and data. Specifically, the modeling review should include:

  (a) Techniques used to collect data on the geology, hydrology, geochemistry, processes, plume geometry/extent

  (b) Technical bases for evaluating effects of the geology, hydrology, geochemistry, processes, and physical phenomena on flow and transport pathways

  (c) Consistency and adequacy of model assumptions

  (d) Technical bases for the concentrations of contaminants

  (e) Sufficiency of data and selection of model parameters and simplifications

  (f) Evaluation of uncertainty associated with model parameters

  (g) Model results compared to more detailed model results or site data (i.e., model validation)

(2) Estimates of the concentrations and lateral and vertical dispersion of those chemicals that may persist in leached-out well field production zones after termination of *in situ* leaching operations and before restoration activities.

(3) Descriptions of proposed methods and techniques to be used to restore ground-water quality, including identification of *in situ* chemical reactions that may hinder or enhance restoration.

(4) A schedule for sequential restoration of well fields.

(5) Descriptions of the expected post-reclamation conditions and quality of restored ground waters, compared with the pre-operational water quality characteristics, and any prior experience restoring ground water at the site.

(6) Adverse effects of the proposed water quality restoration operations on ground waters outside production zones.

(7) Procedures to be used for plugging, sealing, capping, and abandoning wells.

(8) Methods of effluent disposal, such as deep-well injection, discharge to surface water, and land application.

6.1.2   Review Procedures

The staff should review plans and schedules for ground-water quality restoration, and perform the following actions:

(1)   If numerical ground-water flow or transport modeling is used to support or develop the ground-water restoration plans, examine the descriptions of features, physical phenomena, and the geological, hydrological, and geochemical aspects of the modeled aquifers. The staff should verify that the descriptions are adequate and that the conditions and assumptions used in the modeling are realistic or reasonably conservative and supported by the body of data presented in the descriptions.

Evaluate the sufficiency of data used to support model input parameter values. Data sources may include a combination of techniques such as laboratory experiments, aquifer hydraulic testing and water level measurements in wells, geochemical analyses, or other site-specific field measurements.

Evaluate the technical bases for parameter ranges, probability distributions, or bounding values. The reviewer should determine whether the parameter values are derived from either site-specific data, or an analysis to show assumed parameter values bound data uncertainty in a manner that is not overly optimistic.

Evaluate whether there are aspects of the model where additional data could provide new information that could invalidate the modeling results and significantly affect the ground-water restoration plan. For example, if constant head boundary conditions are used in a numerical ground-water flow model, could additional wells or sampling during a different season result in a significantly different interpretation of model boundary conditions? If so, is a different interpretation of boundary conditions likely to significantly alter model results used to develop or support the restoration plan?

Examine the initial conditions and boundary conditions used in any numerical modeling for consistency with available data. The staff should also consider the potential importance of temporal and spatial variations in boundary conditions and source terms used to support the ground-water restoration plan.

Evaluate the applicant's assessment of uncertainty and variability in model parameters. The reviewer should determine whether uncertainty in both temporal and spatial parameter variability is incorporated into or bounded by parameter values.

Examine the technical bases for the identification of post-extraction changes to ground-water quality. The staff should examine how the evolution of water quality has been incorporated into estimates of restoration time or the number of pore volumes required to attain restoration goals.

Examine the assumptions used to develop any model of reactive transport that accounts for site geochemical processes, such as sorption or any other geochemical reaction, that reduce concentrations of, or retard, contaminants. The modeling should consider available data about the native ground-water downgradient of the production areas, the geochemical environment, hydraulic and transport properties, and the spatial variations of aquifer properties and ground-water volumetric fluxes along the flow paths.

Evaluate the estimated restoration time or required number of pore volume displacements for consistency with the output from any numerical model of ground-water restoration.

The reviewer should evaluate whether the applicant has appropriately reduced the dimensionality and complexity of models. The dimensionality of models, heterogeneity of aquifer parameters, and significant process couplings may be reduced if it is shown that the reduced and simplified dimension model bounds the prediction of the full dimension model. The staff should evaluate the acceptability of the sensitivity analyses used to support the model of the ground-water restoration and the estimation of restoration time and pore volume displacements.

Where appropriate, the reviewer may use an alternative model to perform an independent technical assessment of ground-water restoration.

(2)   Evaluate estimates of post-extraction ground-water quality by comparison to descriptions of lixiviant composition and host rock geochemistry. Ensure that methods for estimating the affected pore volume are consistent with the methods used at any research and development site or other sites upon which restoration estimates may be based.

(3)   Compare descriptions of the proposed restoration methods with those methods that have been successfully applied at other *in situ* leaching facilities. Sources of information can include research and development and production sites that are located in similar hydrogeologic environments and have used similar restoration techniques. However, the applicant is not required to present operational experience from a research and development facility as part of an application. Ensure that the proposed restoration methods are appropriate for the host rock and lixiviant chemistry.

(4)   Assess whether the applicant has provided a reasonable standard for the determination of restoration success and a realistic assessment of the expected post-restoration water quality by comparing standards with previous restoration work at the research and development site or other previously restored *in situ* leaching facilities.

(5)   Evaluate the ability of the post-reclamation stability monitoring program to verify successful restoration.

(6)   Consider whether the proposed restoration program adequately addresses water quality cleanup because of well field flare (undetected spread of extraction solutions between the well field and monitor wells of the production zone), and whether the quantity of water pumped during restoration will adversely affect off-site ground-water uses.

(7)   Assess whether plans for plugging and abandoning wells before license termination are consistent with generally accepted techniques.

(8)   Assess whether plans for methods of effluent disposal, such as deep-well injection, discharge to surface water, and land application are consistent with generally accepted techniques.

For license renewals and amendment applications, Appendix A to this standard review plan provides guidance for examining facility operations and the approach that should be used in evaluating amendments and renewal applications.

6.1.3 Acceptance Criteria

The primary purpose of restoring the ground-water quality in a well field after the completion of uranium extraction operations is to assure the protection of public health and the environment. NRC shares the regulatory oversight of ground-water restoration with the EPA under its Underground Injection Control Program (40 CFR Part 144) and those underground injection control programs administered by EPA Authorized States. In addition to the NRC license, the EPA Authorized States issue underground injection control permits for *in situ* leaching operations, after the EPA grants an exemption from ground-water protection provisions for the portion of the aquifer undergoing uranium extraction (the exploited ore zone in an aquifer). The EPA aquifer exemption effectively removes that portion of the aquifer from any future consideration for ground-water protection; however, the ground-water protection provisions are still in effect for the aquifer adjacent to the exempted area. The EPA Authorized State may impose ground-water restoration requirements that are more stringent than the delegated federal program. Ground-water restoration requirements may vary from state to state. The reviewer is advised to closely coordinate the NRC licensing review activities with the underground injection control permitting programs of EPA Authorized States to avoid unnecessary duplication of effort. The following acceptance criteria should serve as the minimum requirements for demonstrating acceptability for the NRC licensing review.

The plans and schedules for ground-water quality restoration are acceptable if they meet the following criteria:

(1)   The application includes estimates of the volume and quality of extraction solutions that need to be cleaned up during ground-water restoration. Generally, these estimates may be based on either experience with previous *in situ* leach operations or research and development investigations in similar host rock. Documentation of such prior experience should be included or referenced in the application. The applicant may also use numerical or analytical ground-water flow and transport modeling to support development of the ground-water restoration plan. When flow and transport modeling is used, the applicant must provide data and model justification to demonstrate that conclusions used to develop the restoration plan are reasonable. Data and model justification must meet the following criteria.

Important design features, physical phenomena, and consistent and appropriate assumptions are identified and described sufficiently for incorporation into the modeling that supports the ground-water restoration plan.

The applicant provides sufficient data to justify the selection of models used to develop the ground-water restoration plan and to adequately define model parameters, initial and boundary conditions, and any simplifying assumptions.

Parameter values, assumed ranges, probability distributions, and/or bounding assumptions used in modeling ground-water restoration are technically defensible and reasonably account for uncertainties and variabilities. The technical bases for each parameter value, ranges of values, or probability distributions used in the modeling ground-water restoration are provided.

In the case of sparse data and/or low confidence in the quality of available data or parameter estimates, the applicant demonstrates by sensitivity analyses or other methods that the proposed ground-water restoration approach is appropriate, and the contingency built into the surety is consistent with the uncertainties.

For reactive transport models, adequate site geochemical data are provided to support the ground-water restoration plans and models. Water chemistry data are needed to develop an understanding of geochemical evolution as ground water is restored in the subsurface. The important geochemical parameters that should be delineated include pH, Eh, dissolved oxygen, temperature, major cation and anion concentrations, concentrations of potential contaminants, and host-rock mineralogy.

Reactive transport models incorporate thermodynamic data on solid phases and aqueous species, allowing the mass action calculations that determine estimated aqueous concentrations and solid phase evolution. Thermodynamic parameters constitute a major source of uncertainty in geochemical modeling, with potentially large effects on predicted aqueous ion concentrations. Therefore, geochemical modeling supporting ground-water restorations should include sensitivity analyses that provide assurance that contaminant concentrations will not be underestimated. Likewise, any kinetic models employed are subjected to critical analysis because of the large influence of kinetic effects at low temperatures. Additionally, consideration of geochemical model limitations and their effects on uncertainty is an important component of the review by the NRC. Such limitations include: the assumption of local equilibrium, neglect of porosity changes caused by precipitation or dissolution of the solid phase, omitting colloidal transport; neglect of density effects due to varying total dissolved solids, simplifying the mineralogical suite, and neglecting surface reactions such as ion exchange.

The applicant documents how the model output is validated in relation to site characteristics.

(2)   The applicant describes the method used for estimating well field *pore volume*[1] and the
        associated horizontal and vertical *flare*.[2]

        A pore volume is an indirect measurement of a unit volume of aquifer water affected by
        in situ leach extraction. It represents the volume of water that fills the void space inside
        a certain volume of rock or sediment. Typically, a pore volume is calculated by
        multiplying the surficial area of a well field (the area covered by injection and recovery
        wells) by the thickness of the production zone being exploited and the estimated or
        measured porosity of the aquifer material. The horizontal and vertical flares are usually
        expressed as additional percentages that are multiplied to the calculated pore volume.
        Specific flare factors approved in the past vary from 20 to 80 percent and are typically
        based on experience from research and development pilot demonstrations. The pore
        volume and flare factors provide a means of comparing the level of effort required to
        restore ground water regardless of the scale of the test. In general, the more pore
        volumes of water it takes to restore ground-water quality, the more effort it will cost to
        achieve restoration.

(3)   The application includes well field restoration plans.

        Restoration plans contain descriptions of the process to be used for well field restoration
        and projected completion schedules. This description should include restoration flow
        circuits, treatment methods, methods for disposal or treatment of wastes and effluents,
        monitoring schedules, a discussion of chemical additives used in the restoration
        process, anticipated effects of chemical additives, and alternate techniques that may be
        employed in the event that primary plans are not effective. Typically, restoration is
        divided into distinct sequential phases in which different techniques are employed.
        Ground-water sweep is used to pump water from the ore zone without reinjecting, to
        recall lixiviant from the aquifer and draw in surrounding uncontaminated water. Reverse
        osmosis/permeate injection circulates water from the well field through a reverse
        osmosis treatment process and reinjects the permeate into the well field, typically at
        rates similar to those used during production. Ground-water recirculation is used to
        evenly distribute water throughout the restored well field, to dilute any pockets of
        remaining contamination. An additional acceptable restoration method is the injection of
        chemical reductants (usually hydrogen sulfide, sodium sulfide, or sodium bisulfide) into
        the well field. These reductants are used to immobilize metals that may have been

---

[1]*Pore volume* is a term of convenience used by the *in situ* leach industry to describe the quantity of free water in the
pores of a given volume of aquifer material. It provides a unit reference that an operator can use to describe the
amount of lixiviant circulation needed to leach an ore body, or describe the unit number of treated water circulations
needed to flow through a depleted ore body to achieve restoration. A pore volume provides a way for an operator to
use relatively small-scale studies and scale the results to field-level pilot tests or to commercial well field scales.

[2]*Flare* is a proportionality factor designed to estimate the amount of aquifer water outside of the pore volume that has
been impacted by lixiviant flow during the extraction phase. The flare is usually expressed as a horizontal and
vertical component to account for differences between the horizontal and vertical hydraulic conductivity of an
aquifer material.

dissolved by the oxidizing lixiviant; however, some general water quality parameters, such as total dissolved solids, may be adversely affected by reductants.

NRC allows flexibility and innovation in approaches to restoration. Therefore, applicants are not limited to one restoration method for all well fields. Rather, they should describe the sequential phases of restoration that may be used and the most likely restoration scenario, based on research and development results and restoration experience. Other restoration approaches, such as in-place biological remediation techniques, have been discussed by some applicants. These techniques show promise, but have not been tested or evaluated at commercial scale *in situ* leach operations. The application of other restoration techniques may necessitate some form of pilot demonstration to evaluate the potential for unanticipated impacts, such as clogging of aquifer pore spaces or potential health impacts from introduced compounds and organisms, before the techniques are applied to full-scale operations.

Restoration plans should also include a list of monitored constituents, a monitoring interval, and the sampling density (wells/acre). An acceptable constituent list should be based on the chemistry of the production and restoration solutions used and on the host rock geochemistry. In the interest of minimizing expense, the applicant may propose a limited set of indicator constituents to monitor restoration progress and a sampling density that does not include all production and injection wells. The applicant may also propose monitoring composite samples from the restoration stream. However, all wells that were sampled for baseline conditions should be sampled for the full list of monitored constituents before a determination of restoration success is made.

The applicant should specify the criteria that will be used to determine restoration success. Generally, the acceptance criteria for restoration success are based on the ability to meet the predetermined numerical standards of the restoration program and the absence of a significant increasing trends of monitored indicator constituent concentrations during the stability monitoring period.

For purposes of surety bonding, restoration plans must include estimates of the level of effort (typically in terms of pore volume displacements) necessary to achieve the primary restoration target concentrations. These estimations may be based on historical results obtained from the research and development site or experience in other well fields having similar hydrologic and geochemical characteristics.

(4)   Restoration standards are established in the application for each of the monitored constituents.

The applicant has the option of determining numerical restoration limits for each monitored constituent on a well-by-well basis, or as a statistical average applied over the entire well field. Restoration standards must be established for the production zone and for any overlying or underlying aquifers that have the potential to be affected by *in situ* leach solutions.

(a)   Primary Restoration Standards—The primary goal of a restoration program is to
        return the water quality within the exploited production zone and any affected
        aquifers to pre-operational (baseline) water quality conditions. Recognizing that
        in situ leach operations fundamentally alter ground-water geochemistry,
        restoration activities are not likely to return ground-water quality to exact water
        quality that existed at every location prior to *in situ* leach operations. Still, as a
        primary restoration goal, licensees are required to attempt to return the
        concentrations of the monitored water quality indicator constituents to within the
        baseline range of statistical variability for each constituent. This standard
        requires licensees to identify the type of statistical analysis and criteria that will
        be used to determine whether concentrations of water quality parameters in the
        affected aquifers fall within an acceptable range of baseline variability. Statistical
        approaches for determining whether contamination persists in affected aquifers
        are found in American Society for Testing and Materials Standard D 6312
        (American Society for Testing and Materials, 2001).

(b)   Secondary Restoration Standards—*In situ* leach operations may cause
        permanent changes in water quality within the exploited production zone,
        because the *in situ* leach extraction process relies on changing the chemistry in
        the production zone to remove the uranium. The applicant may therefore
        propose returning the water quality to its pre-operational class of use
        (e.g., drinking water, livestock, agricultural, or limited use) as a secondary
        restoration standard. Applications should state the principal goal of the
        restoration program and that secondary standards will not be applied so long as
        restoration continues to result in significant improvement in ground-water quality.
        The applicant must first attempt to return ground-water quality to primary
        restoration standards before falling back on secondary restoration standards.
        License conditions should be set up such that a license amendment is necessary
        before the applicant can revert to secondary goals. The applicant must commit
        to use reasonable efforts to reach primary restoration standards.

        It is acceptable to establish secondary restoration standards on a constituent-by-
        constituent basis, with the numerical limits established to ensure state or EPA
        primary or secondary drinking water standards will not be exceeded in any
        potential source of drinking water. For radionuclides not included in the drinking
        water standards, it is acceptable to determine, on a constituent-by-constituent
        basis, secondary standards from the concentrations for unrestricted release to
        the public in water, from Table 2 of 10 CFR Part 20, Appendix B.

(c)   If a constituent cannot technically or economically be restored to its secondary
        standard within the exploited production zone, an applicant must demonstrate
        that leaving the constituent at the higher concentration would not be a threat to
        public health and safety or the environment or produce an unacceptable
        degradation to the water use of adjacent ground-water resources. This situation
        might arise with respect to general water quality parameters such as the total
        dissolved solids, sulfate, chloride, iron, and others which do not typically present

a health risk. However, not all the major constituents have a primary or
secondary drinking water standard (e.g., bicarbonate, carbonate, calcium,
magnesium, and potassium). Consequently, ground-water restoration may
achieve the secondary standard for total dissolved solids, but may not achieve a
secondary standard for individual major ions that contribute to total dissolved
solids. If such a situation occurred, the applicant must show that leaving the
individual constituent at a concentration higher than secondary standard would
not be a threat to public health and safety nor the environment or produce an
unacceptable degradation to the water use of adjacent ground-water resources.
Such proposed alternatives must be evaluated on a case-by-case basis as a
license amendment request only after restoration to the primary or secondary
standard is shown not to be technically or economically achievable. This
approach is consistent with the as low as is reasonably achievable philosophy
that is used broadly within NRC.

(5)   The post-restoration stability monitoring program is described in the application.

The purpose of a stability monitoring program is to ensure that chemical species of
concern do not increase in concentration subsequent to restoration. The applicant
should specify the length of time that stability monitoring will be conducted, the number
of wells to be monitored, the chemical indicators to be monitored, and the monitoring
frequency. These requirements will vary based on site-specific post-extraction water
quality and geohydrologic and geochemical characteristics. Before final well field
decommissioning is completed, all designated monitor wells must be sampled for all
monitored constituents. Well fields may be decommissioned when all constituent
concentrations meet approved restoration standards and no post-restoration
degradation in ground-water quality occurs outside of the aquifer exemption boundary.

(6)   The application includes a discussion of the likely external effects of
ground-water restoration.

Ground-water restoration operations, and the expected post-reclamation ground-water
quality, must not adversely affect ground-water use outside the exploited production
zone. Water users from nearby municipal or domestic wells that were in use before
in situ leach operations should be provided reasonable assurance that their water quality
will not be impacted. Impacts are not limited to chemical constituent concentrations, but
also include changes in color, odor, hardness, and taste of the water. The water quality
outside the exploited production zone should not, as a result of *in situ* leach operations,
exceed EPA primary or secondary drinking water standards for ground water.
Ground-water quality should not exceed the appropriate state water-use standards for
aquifers that cannot support a drinking water use.

(7)   Methods for abandoning wells are included in the application.

The basic purpose for sealing abandoned wells and bore holes is to restore the well field to pre-operational hydrogeologic conditions. Any well or bore hole to be permanently abandoned should be completely filled in such a manner that vertical movement of water along the borehole is prevented. *In situ* leach operators usually rely on a drilling contractor to perform well abandonment. The application should specify the methods and materials to be used to plug holes, and that records documenting the well abandonment will be maintained by the licensee. Abandonment procedures that: (i) conform to American Society for Testing and Materials Standard D 5299 (1992); (ii) are from the State Engineer's Office; or (iii) are codified in state regulations or rules are considered acceptable. An applicant may propose other generally accepted standards for abandoning wells and boreholes. References for these standards should be specified in the application and copies should be kept on file by the applicant. Techniques proposed by the applicant that are not considered to be generally accepted abandonment practices should be described in detail and may require additional time for review.

(8)   Descriptions of water consumption impacts.

During *in situ* leach operations, water quality impacts usually are more of a concern than water consumption impacts. This is because water consumption during *in situ* leach operations is relatively small. However, when restoration activities begin, water consumption may significantly increase. The amount of increase will depend on the restoration techniques applied. Techniques that clean up the aquifer by pumping water from the aquifer, cleaning the water, and reinjecting the clean water consume the least amount of water. Water consumption impacts will result in water loss from the aquifer and water level declines. The impacts of water consumption on local wells and water users should be evaluated. Water level declines can result in increased pumping costs or inability to obtain water from the aquifer in local wells. Water loss from the aquifer may mean that less water could be available to down gradient ground-water and surface-water users.

(9)   The applicant may propose alternatives to restoring an exploited production zone to primary or secondary ground-water restoration standards in lieu of the above criteria. These alternatives must be evaluated on a case-by-case basis and must assure protection of human health and the environment and assure no unacceptable degradation to adjacent ground-water resources. As an example, if an applicant proposes no ground-water restoration activities within the exploited production zone, the applicant would be required to show that adequate institutional control provisions are in place to assure potential water supplies adjacent to the exploited production zone would not be accessed for a use that would harm human health or the environment. If predictive computer modeling is used to support this alternative, the model must be validated by comparing the modeling results to ground-water monitoring for an appropriate period of time after *in situ* leach operations cease in a well field. The applicant must maintain a financial surety to cover potential restoration costs in the event that monitoring results are contrary to model predictions and ground-water restoration must be initiated.

(10)   Onsite Evaporation

Liquid waste and solid wastes (sludge) from surface impoundments resulting from *in situ*
leach operations are 11 e.(2) byproduct material. Licensees must demonstrate that
surface impoundments are designed, operated, and decommissioned in a manner that
prevents migration of waste from the surface impoundment to subsurface soil, ground
water, or surface water in accordance with 10 CFR Part 40, Appendix A. Applicants
must also demonstrate that monitoring requirements are adequately established to
detect any migration of contaminants to the ground water. Solid waste material must be
disposed of in an existing tailings impoundment or 11e.(2) disposal cell in accordance
with 10 CFR Part 40, Appendix A, Criterion 2.

Surface impoundments will be found acceptable if they comply with the design
provisions for surface impoundments [Criteria 5A(1) through 5A(5)]; installation of liners
and leak detection (Criterion 5E); seepage control (Criterion 5F); and radium cleanup
standards [Criterion 6(6)] of 10 CFR Part 40, Appendix A.

(11)   Release In Surface Waters

Process waste water resulting from *in situ* leach operations is 11e.(2) byproduct
material. The U.S. Environmental Protection Agency (EPA), in accordance with
40 CFR 440.34, does not allow new ISL facilities to discharge process waste water to
navigable waters. For release of this waste to surface waters, existing licensees must
meet the requirements of 10 CFR 20.1302(b)(2), and should demonstrate that doses
are maintained as low as is reasonably achievable (ALARA). NRC has no specific
requirements for non-radiological constituents, and may adopt the appropriate State
limits. Anticipated discharge must be described in enough detail to evaluate
environmental impacts. Appropriate State and Federal agency permits should be
obtained in accordance with 10 CFR 20.2007.

(12)   Land Applications

For the land application of process waste water, the applicant must meet the regulatory
provisions in 10 CFR 20.2002 and demonstrate that doses are maintained ALARA
within the dose limits in 10 CFR 20.1301. Proposed land application activities should be
described in sufficient detail to satisfy the NRC need to assess environmental impacts.
This may require analysis to assess the chemical toxicity of radioactive and
nonradioactive constituents. Specifically, licensees must provide: (i) a description of the
waste, including its physical and chemical properties that are important to risk
evaluation; (ii) the proposed manner and conditions of waste disposal; (iii) projected
concentrations of radioactive contaminants in the soil; and (iv) projected impacts on
ground-water and surface-water quality and on land uses, especially crops and
vegetation. In addition, projected exposures and health risks that may be associated
with radioactive constituents reaching the food chain must be analyzed to ensure that
doses are ALARA. Proposals should include provisions for periodic soil surveys to verify
that contaminant levels in the soil do not exceed those projected, and should also

include a remediation plan that can be implemented if projected levels are exceeded. Appropriate State and Federal agency permits must be obtained in accordance with 10 CFR 20.2007. The applicant must also comply with NRC regulatory provisions for decommissioning. The applicant should also address whether the proposed land applications methodologies will comply with 10 CFR Part 40, Appendix A, Criterion 6(6), at the time of decommissioning.

(13)   Deep-Well Injection

Proposals for disposal of liquid waste from process water by injection in deep wells must meet the regulatory provisions in 10 CFR 20.2002 and demonstrate that doses are ALARA and within the dose limits in 10 CFR 20.1301. The injection facility should be described in sufficient detail to satisfy the NRC need to assess environmental impacts. Specifically, proposals must include: (i) a description of the waste, including its physical and chemical properties important to risk evaluation; (ii) the proposed manner and conditions of waste disposal; (iii) an analysis and evaluation of pertinent information on the nature of the environment; (iv) information on the nature and location of other potentially affected facilities; and (v) analyses and procedures to ensure that doses are ALARA, and within the dose limits in 10 CFR 20.1301.

In addition, pursuant to the provisions of 10 CFR 20.200, proposals for disposal by injection in deep wells should also meet any other applicable Federal, State, and local government regulations pertaining to deep well injection. Applicants must obtain any necessary permits for this purpose. In particular, proposals must satisfy the EPA regulatory provisions in 40 CFR Part 146: Underground Injection Control (UIC) Program: Criteria and Standards, and applicants must obtain necessary permits from EPA and/or States authorized by EPA to enforce these provisions. In general, applications that satisfy EPA regulations under the UIC Program, which are approved by the EPA or an EPA-authorized State issuing the UIC permit and the applicable provisions of 10 CFR Part 20, will also be approved by the staff. Licensees and applicants disposing of liquid waste from process water by injection in deep wells are further required to comply with NRC regulatory provisions for decommissioning.

6.1.4   Evaluation Findings

If the staff review, as described in this section, results in the acceptance of the plans and schedules for ground-water quality restoration, the following conclusions may be presented in the technical evaluation report.

NRC has completed its review of the plans and schedules for ground-water quality restoration proposed for use at the _____ *in situ* leach facility. This review included an evaluation of the methods that will be used to develop the ground-water restoration program and schedules using the review procedures in standard review plan Section 6.1.2 and the acceptance criteria outlined in standard review plan Section 6.1.3.

Ground-Water Quality Restoration, Surface
Reclamation, and Plant Decommissioning

The applicant has committed to adopt well field ground-water restoration standards that are representative of the pre-operational baseline ground-water conditions. As a secondary restoration goal, the applicant has identified and committed to ensure federal or state drinking water standards will not be exceeded outside of the aquifer exemption boundary as a result of operations.

The applicant's method for estimating well field pore volume is acceptable, taking into account the estimated effective porosity of the contaminated region and the lateral and vertical extent of contamination. With respect to the methodology for undertaking restoration, the applicant provided an acceptable approach that includes a mix of ground-water sweep, reverse osmosis, and ground-water recirculation. The well-field-specific mix of these approaches will be determined as part of the ground-water restoration plan for each individual well field. In addition, the applicant has proposed an acceptable method for determining the extent of well field flare and for ensuring acceptable restoration of the flare. The applicant has committed to an acceptable schedule for complete restoration for any well field after ore extraction ceases.

The applicant has presented an acceptable list of indicator constituents to be monitored and has specified acceptable criteria to determine the success of restoration either on a well-by-well or well field average basis. The number of pore volume replacements necessary to achieve the primary restoration targets has been provided and is acceptable. The applicant has adopted a primary restoration program that will return the water quality of the production zone and affected aquifers to pre-extraction (baseline) water quality, that any secondary restoration standards proposed by the applicant are acceptable, or that final water quality will protect public health and safety and the environment in compliance with as low as is reasonably achievable principles. The applicant's post-restoration stability monitoring program is acceptable.

The methods proposed for abandoning wells and sealing them to restore the well field to pre-extraction hydrologic conditions are acceptable. The applicant has evaluated the consumptive water impacts of the *in situ* leach facility using acceptable methods.

Based on the information provided in the application and the detailed review conducted of the plans and schedules for ground-water quality restoration for the _____ *in situ* leach facility, the staff concludes that the proposed plans and schedules for ground-water quality restoration are acceptable and are in compliance with 10 CFR 40.32(c), requiring the applicant's proposed equipment, facilities, and procedures to be adequate to protect health and minimize danger to life or property; 10 CFR 40.32(d), requiring that the issuance of the license will not be adverse to the common defense and security or to the health and safety of the public; and 10 CFR 51.45(c), which requires the applicant to provide sufficient data for the Commission to conduct an independent analysis. The related reviews of the 10 CFR Part 51 environmental protection regulations for domestic licensing and related regulatory functions for plans and schedules for ground-water restoration in accordance with standard review plan Sections 5.0, "Operations;" and 7.0, "Environmental Effects;" are addressed elsewhere in this technical evaluation report.

6.1.5   References

American Society for Testing and Materials. "Standard Guide for Developing Appropriate Statistical Approaches for Ground-Water Detection Monitoring Programs, Designation: D6312." West Conshohocken, Pennsylvania: American Society for Testing and Materials. 2001.

─────. "Standard Guide for Decommissioning of Ground Water Wells, Vadose Zone Monitoring Devices, and Other Devices for Environmental Activities, Designation: D 5299." West Conshohocken, Pennsylvania: American Society for Testing and Materials. 1992.

6.2   Plans For Reclaiming Disturbed Lands

6.2.1   Areas of Review

Prior to commencement of reclamation, the licensee will provide the NRC with maps and data that document the post-operational condition. The staff should also review plans for (i) reclaiming temporary diversion ditches and impoundments; (ii) reestablishing surface drainage patterns disrupted by the proposed activities; and (iii) returning the ground surface and structures for post-operational use (i.e., license termination), in accordance with the criteria in Section 6.4 of the standard review plan.

Staff should review the pre-remediation radiological survey program that will identify areas of the site that need to be cleaned up to comply with NRC concentration limits. The staff should evaluate measurement techniques and sampling procedures proposed for determining the radionuclide concentrations and the extent of contamination of structures and soils. In addition, the review should confirm that the licensee will have an approved decommissioning radiation protection program in place before the start of reclamation and cleanup work and that an acceptable agreement is in place for off-site disposal of 11e.(2) byproduct material.

6.2.2   Review Procedures

The staff should determine whether the described approaches for reclaiming temporary diversion ditches and impoundments, reestablishing surface drainage patterns disrupted by the proposed activities, and returning the ground surface and structures for post-operational use are consistent with regulatory guidance and are sufficient to satisfy the requirements of 10 CFR Part 40, Appendix A, Criterion 6(6), and 10 CFR 40.42. The staff should ensure that the licensee intends to restore topography and vegetation to a state that is similar to pre-operational conditions. The staff should review the pre-reclamation survey plan to ensure that it provides adequate coverage to designate contaminated areas for cleanup. Particular attention should be focused on sampling temporary diversion ditches and surface impoundments, well field surfaces, process and storage areas, transportation routes, and operational air monitoring locations. These areas are expected to have higher levels of contamination than surrounding areas. The staff should also ensure that plans exist for the disposal of contaminated soils at an existing licensed byproduct material disposal facility, consistent with 10 CFR Part 40, Appendix A, Criterion 2. The staff should confirm that the licensee has an approved radiological protection program to ensure worker safety during decommissioning, reclamation, and cleanup activities. Prior to commencement of reclamation, the NRC should review licensee commitments and any changes the licensee has proposed. The program for radiation protection is addressed in Section 5.7 of the standard review plan but additional review is needed to ensure any hazards specific to decommissioning are addressed (e.g., yellowcake dryer demolition). The staff should review the compliance history for the radiation safety program to identify any deficient areas that may require special consideration before the start of work.

For license renewals and amendment applications, Appendix A to this standard review plan provides guidance for examining facility operations and the approach that should be used in evaluating amendments and renewal applications.

6.2.3 Acceptance Criteria

The plans for reclaiming disturbed lands are acceptable if they meet the following criteria:

(1) Appropriate cleanup criteria will be used in conducting the pre-reclamation surveys and planned cleanup activities. Acceptable cleanup criteria are discussed in standard review plan Sections 6.3 (for structures) and 6.4 (for soils).

(2) The pre-reclamation radiological survey program for buildings and soils identifies instruments and techniques similar to those used in the pre-operational survey program to determine baseline site conditions (e.g., background radioactivity) but also takes into account current technology (acceptable sensitivity), results from operational monitoring, and other information that provide insights to areas of expected contamination.

Survey areas should include diversion ditches, surface impoundments, well field surfaces and structures in process and storage areas, on-site transportation routes for contaminated material and equipment, and other areas likely to be contaminated. A sampling grid of 100 m^2 (for soil) should be used and a statistical basis for sample size should be provided. Acceptable methods for sampling are provided in NUREG–1575, "Multi-Agency Radiation Survey and Site Investigation Manual (MARSSIM)" (NRC, 2000).

(3) The licensee provides the procedures for interpretation of the pre-reclamation survey results and describes how they will be used to identify candidate areas for cleanup operations. Acceptable survey methods are discussed in standard review plan Section 6.4, "Procedures for Conducting Post-Reclamation and Decommissioning Radiological Surveys."

(4) The discussion of surface restoration includes a pre-construction surface contour map, a description of any significant disruptions to surface features during facility construction and operation, and a description of planned activities for surface restoration that identifies any important features that cannot be restored to the pre-operations condition.

(5) Any changes to the existing NRC-approved radiation safety program that are needed for decommissioning and reclamation work are identified with appropriate justification to assure continued safety for workers and the public. Acceptable approaches for the

radiation safety program are evaluated in accordance with Section 5.7 of this standard review plan, "Radiation Safety Controls and Monitoring."

(6)   The applicant has an approved waste disposal agreement for 11e.(2) byproduct material disposal at an NRC or NRC Agreement State licensed disposal facility. This agreement is maintained on site. The applicant has committed to notify NRC in writing within 7 days if this agreement expires or is terminated and to submit a new agreement for NRC approval within 90 days of the expiration or termination. Failure to comply with this license condition will result in a prohibition from further lixiviant injection.

(7)   The applicant commits to providing final (detailed) reclamation plans for land (soil) to the NRC for review and approval at least 12 months before the planned commencement of reclamation of a well field or licensed area. The final decommissioning plan includes a description of the areas to be reclaimed, a description of planned reclamation activities, a description of methods to be used to ensure protection of workers and the environment against radiation hazards.

(8)   The decommissioning plan addresses the non-radiological hazardous constituents associated with the wastes according to 10 CFR Part 40, Appendix A, Criterion 6(7). Any unusual or extenuating circumstances related to such constituents should be discussed in the reclamation plan or decommissioning plan in relation to protection of public health and the environment and should be evaluated by staff.

(9)   The quality assurance and quality control programs address all aspects of decommissioning. The programs should indicate a confidence interval or that one will be specified before collection of samples. The data to be used to demonstrate compliance and the quality assurance procedures to confirm that compliance data are precise and accurate are identified. Management will ensure that approved procedures are followed.

6.2.4 Evaluation Findings

If the staff review, as described in this section, results in the acceptance of the plans for reclaiming disturbed lands, the following conclusions may be presented in the technical evaluation report.

NRC has completed its review of the plans for reclaiming disturbed lands proposed for use at the _____ in situ leach facility. This review included an evaluation of the methods that will be used to develop the reclamation of disturbed lands program using the review procedures in standard review plan Section 6.2.2 and the acceptance criteria outlined in standard review plan Section 6.2.3.

The applicant has acceptable plans for a pre-reclamation radiation survey that use instrumentation and techniques similar to the pre-operational survey used to establish baseline site conditions, if these are still acceptable methods. The applicant has acceptably considered results from operational monitoring and other information relative to areas of expected contamination in its reclamation plans. Areas to be evaluated include diversion ditches, surface impoundments, well field surfaces, and structures in process and storage areas, on-site transportation routes, and other areas likely to be contaminated. The applicant has proposed acceptable methodology to determine areas to be resampled or sampled with higher than normal densities. The applicant has defined appropriate procedures for the pre-reclamation survey and the means used to identify areas for cleanup using the acquired data. Methods proposed for reclamation and an acceptable plan for surface restoration, including identification of any irreversible changes, have been provided. The applicant has assured NRC that any required changes to the radiation safety program identified as a result of the reclamation work will be implemented before commencing the work.

Based on the information provided in the application and the detailed review conducted of the plans for reclaiming disturbed lands for the _____ in situ leach facility, the staff concludes that the proposed plans are acceptable and are in compliance with 10 CFR 40.32(c), which requires applicant proposed equipment, facilities, and procedures to be adequate to protect health and minimize danger to life or property; 10 CFR 40.42(g)(4), which provides requirements for final decommissioning plans; 10 CFR 40.41(c), which requires the applicant to confine source or byproduct material to the locations and purposes authorized in the license; 10 CFR Part 40, Appendix A, Criterion 2, which requires that the applicant provide objective evidence of an agreement for disposal of 11e.(2) byproduct materials either in a licensed waste disposal site or at a licensed mill tailings facility to demonstrate non-proliferation of waste disposal sites; 10 CFR Part 40, Appendix A, Criterion 6(6), which identifies cleanup criteria requirements; and 10 CFR 51.45(c), which requires the applicant to provide sufficient data for the Commission to conduct an independent analysis.

The reclamation plan specifies the location of records of information important to the decommissioning as required by 10 CFR 40.36(f) and meets the criteria of 10 CFR 40.42(g)(4) and (5). The plan sufficiently demonstrates that the proposed reclamation activities will result in compliance with 10 CFR 40.42(j)(2) requirements to conduct a radiation survey. The plan complies with the 10 CFR 40.42(k)(1) and (2) requirements that source material be properly disposed of and reasonable effort be made to eliminate residual radioactive contamination. The plan demonstrates the proposed reclamation activities will result in compliance with 10 CFR Part 40, Appendix A, Criterion 6(7) requirements to prevent threats to human health and the environment from non-radiological hazards.

6.2.5 Reference

NRC. NUREG–1575, "Multi-Agency Radiation Survey and Site Investigation Manual (MARSSIM)." Revision 1. Washington, DC: NRC. 2000.

6.3   Removal and Disposal of Structures, Waste Materials, and Equipment

6.3.1 Areas of Review

The staff should review methodologies proposed for removal and disposal of contaminated structures and equipment used during in situ leach operations, as well as techniques for managing toxic and radioactive waste materials. The reviewers should also evaluate approaches for identifying radiological hazards before initiating dismantlement of structures and equipment and for detection and cleanup of removable contamination from such structures and equipment. The staff should also review plans for ensuring that all contaminated facilities and equipment are addressed and are either planned to be disposed of in a licensed facility, will meet the contamination levels for unrestricted use, or are designated for re-use at another *in situ* leach facility. The staff should also review provisions made for the removal and disposal of byproduct material to an existing uranium mill or licensed disposal site.

6.3.2 Review Procedures

The staff should determine whether the techniques proposed for removing and disposing of structures and equipment used during *in situ* leach operations and approaches for managing toxic and radioactive waste materials are consistent with regulatory guidance and sufficient to meet the applicable regulatory requirements in 10 CFR 40.42. Plans for structures and equipment to be released for unrestricted use should be reviewed using standard review plan Section 5.7.6, "Contamination Control Program." The staff should confirm that plans for dismantlement of structures and equipment include a preliminary assessment of anticipated hazards that should be considered before dismantlement. This should include the use of appropriate survey methods to determine the extent of contamination of equipment and structures before starting decommissioning and reclamation work. Particular attention should be focused on those parts of the processing system that are likely to have accumulated contamination over long time periods such as pipes, ventilation equipment, effluent control systems, and facilities and equipment used in or near the yellowcake dryer area. The staff should also review provisions made for the removal and disposal of byproduct material to an existing uranium mill or licensed disposal site to ensure that they meet requirements of 10 CFR Part 40, Appendix A, Criterion 2.

For license renewals and amendment applications, Appendix A to this standard review plan provides guidance for examining facility operations and the approach that should be used in evaluating amendments and renewal applications.

6.3.3 Acceptance Criteria

The procedures for removing and disposing of structures, waste materials, and equipment are acceptable if they meet the following criteria:

(1)   A program is in place to control residual contamination on structures and equipment.

(2)   Measurements of radioactivity on the interior surfaces of pipes, drain lines, and duct work will be determined by making measurements at all traps and other appropriate access points, provided that contamination at these locations is likely to be representative of contamination on the interior of the pipes, drain lines, and ductwork.

(3)   Surfaces of premises, equipment, or scrap that are likely to be contaminated but are of such size, construction, or location as to make the surface inaccessible for purposes of measurement are presumed to be contaminated in excess of the limits.

(4)   Before release of structures for unrestricted use, the licensee makes a comprehensive
radiation survey to establish that contamination is within the limits specified in standard
review plan Section 5.7.6, "Contamination Control Program" and obtain NRC approval.

(5)   A contract between the licensee and a waste disposal operator exists to dispose of
11e.(2) byproduct material.

(6)   The applicant commits to providing final (detailed) decommissioning plans for structures
and equipment to the NRC for review and approval at least 12 months before the
planned commencement of decommissioning of such structures and equipment. The
final decommissioning plan includes a description of structures and equipment to be
decommissioned, a description of planned decommissioning activities, a description of
methods to be used to ensure protection of workers and the environment against
radiation hazards, a description of the planned final radiation survey, and an updated
detailed cost estimate. A license condition will be established to this effect.

6.3.4   Evaluation Findings

If the staff review, as described in this section, results in the acceptance of the methodologies
for removal and disposal of structures, waste materials, and equipment, the following
conclusions may be presented in the technical evaluation report.

NRC has completed its review of the methodologies for removal and disposal of structures and
equipment used at the _____ in situ leach facility. This review included an
evaluation of the methods that will be used to develop the procedures for removal and disposal
of structures, waste materials, and equipment using the review procedures in standard review
plan Section 6.3.2 and the acceptance criteria outlined in standard review plan Section 6.3.3.

The applicant has established an acceptable program for the measurement and control of
residual contamination on structures and equipment. The applicant has made acceptable plans
for measurements of radioactivity on the interior surfaces of pipes, drain lines, and ductwork by
making appropriate measurements at all traps and other access points where contamination is
likely to be representative of system-wide contamination. All premises, equipment, or scrap
likely to be contaminated but that cannot be measured, will be assumed by the applicant to be
contaminated in excess of limits and will be treated accordingly. For all premises, equipment,
or scrap contaminated in excess of specified limits, the applicant will provide detailed, specific
information describing the premises, equipment, or scrap in terms of extent and degree of
radiological contamination. The applicant will provide a detailed health and safety analysis that
reflects that the contamination and any use of the premises, equipment, or scrap will not result
in an unacceptable risk to the health and safety of the public or the environment. The applicant
plans to conduct a comprehensive radiation survey to establish that any contamination is within
limits specified before the release of the premises, equipment, or scrap. A contract exists
between the licensee and a licensed waste disposal site operator to dispose 11e.(2) byproduct
material.

Based on the information provided in the application and the detailed review conducted of the
methodologies for removal and disposal of structures, waste materials, and equipment for the
_____ in situ leach facility, the staff concludes that the methodologies are
acceptable and are in compliance with 10 CFR 40.32(c), which provides requirements for final

decommissioning plans; 10 CFR 40.42(g)(4), which requires the applicant's proposed equipment, facilities, and procedures to be adequate to protect health and minimize danger to life or property; 10 CFR 40.41(c), which requires the applicant to confine source or byproduct material to the locations and purposes authorized in the license; and 10 CFR Part 40, Appendix A, Criterion 2, which requires that the applicant provide objective evidence of an agreement for disposal of 11e.(2) byproduct materials either in a licensed waste disposal site or at a licensed mill tailings facility to demonstrate non-proliferation of waste disposal sites.

6.3.5 References

None.

6.4 Methodologies for Conducting Post-Reclamation and Decommissioning Radiological Surveys

6.4.1 Areas of Review

The staff should review methodologies for conducting post-reclamation and decommissioning radiological surveys. The staff should review the radiological verification survey program that will serve as a basis for determining compliance with NRC concentration limits. The staff should evaluate the measurement techniques and sampling procedures proposed.

6.4.2 Review Procedures

The staff should determine whether the methodologies for conducting post-reclamation and decommissioning radiological surveys are acceptable to verify that concentration limits of 10 CFR Part 40, Appendix A, Criterion 6(6) are met. The staff should ensure that sampling and locations are acceptable and representative of conditions at the site. The staff should consider the survey methods provided in NUREG–1575 (NRC, 2000) along with the applicable site conditions to determine the acceptability of the licensee proposed sampling techniques. The staff should confirm that the determination of background concentrations of radium-226 and other radionuclides is based upon sampling in uncontaminated areas near the site. Other radionuclides that should be sampled if suspected to be present include thorium-230, thorium-232, uranium; and lead-210.

The radium benchmark dose applies for cleanup of residual radionuclides other than radium in soil and for surface activity on structures. If appropriate, the reviewer should refer to Appendix E of this standard review plan for guidance on the benchmark approach.

For license renewals and amendment application, Appendix A to this standard review plan provide guidance for examining facility operations and the approach that should be used in evaluating amendments and renewal applications.

6.4.3 Acceptance Criteria

The procedures for conducting post-reclamation and decommissioning radiological surveys are acceptable if they meet the following criteria:

(1)   The cleanup criteria for radium in soils are met as provided in 10 CFR Part 40, Appendix A, Criterion 6(6).

This criterion states that the design requirements for longevity and control of radon releases apply to any portion of a licensed and/or disposal site unless such portion contains a concentration of radium in land, averaged over areas of 100 m², which as a result of byproduct material, does not exceed the background level by more than:

(i)   5 picocuries per gram (pCi/g) of radium-226, or, in the case of thorium byproduct material, radium-228, averaged over the first 15 cm [5.9 in.] below the surface,

(ii)   15 pCi/g of radium-226, or, in the case of thorium byproduct material, radium-228, averaged over 15-cm [5.9-in.] thick layers more than 15 cm [5.9 in.] below the surface.

(2)   Background radionuclide concentrations are determined using appropriate methods as described in Section 2.9, "Background Radiological Characteristics," of this standard review plan. If there are large variations in the background radionuclide concentrations within a given site, the licensee may assign different background radionuclide concentrations to different areas of the site, provided that the licensee properly justifies the background concentrations selected for each area.

(3)   Acceptable cleanup criteria for uranium in soil, such as those in Appendix E of this standard review plan, are proposed by the applicant. This is the radium benchmark dose approach of 10 CFR Part 40, Appendix A, Criterion 6(6).

(4)   For areas that already meet the radium cleanup criteria, but that still have elevated thorium levels, the applicant proposes an acceptable cleanup criterion for thorium-230. One acceptable criterion is a concentration that, combined with the residual concentration of radium-226, would result in the radium concentration (residual and from thorium decay) that would be present in 1,000 years meeting the radium cleanup standard.

(5)   The survey method for verification of soil cleanup is designed to provide 95-percent confidence that the survey units meet the cleanup guidelines. Appropriate statistical tests for analysis of survey data are described in NUREG–1575, "Multi-Agency Radiation Survey and Site Investigation Manual" (NRC, 2000).

6.4.4   Evaluation Findings

If the staff review, as described in this section, results in the acceptance of the methodologies for conducting post-reclamation and decommissioning radiological surveys, the following conclusions may be presented in the technical evaluation report.

NRC has completed its review of the methodologies for conducting post-reclamation and decommissioning radiological surveys proposed for use at the _____ in situ leach facility. This review included an evaluation of the methods that will be used for the post-reclamation and decommissioning radiological surveys using the review procedures in standard review plan Section 6.4.2 and the acceptance criteria outlined in standard review plan Section 6.4.3.

The applicant has developed acceptable methodologies for verification of cleanup (final status survey plan) that demonstrate that the radium concentration in the upper 15 cm [5.9 in.] of soil will not exceed 5 pCi/g and in subsequent 15 cm [5.9 in.] layers will not exceed 15 pCi/g. Also, the cleanup of other residual radionuclides in soil will meet the criteria developed with the radium benchmark dose approach (Appendix E), including a demonstration of ALARA and application of the unity test of 10 CFR Part 40, Appendix A, Criterion 6(6) where applicable. For cases in which the licensee has proposed an alternative to the requirements of Criterion 6(6) or the approved guidance, the staff determines that the resulting level of protection is equivalent to that required by this criterion.

Based on the information provided in the application and the detailed review conducted of the methodologies for conducting post-reclamation and decommissioning radiological surveys for the _____ in situ leach facility, the staff concludes that the methodologies are acceptable and are in compliance with 10 CFR 40.32(c), which requires the applicant's proposed equipment, facilities, and procedures to be adequate to protect health and minimize danger to life or property; 10 CFR 40.32(d), which requires that the issuance of the license will not be inimical to the common defense and security or to the health and safety of the public; 10 CFR 40.41(c), which requires the applicant to confine source or byproduct material to the locations and purposes authorized in the license; 10 CFR Part 40, Appendix A, Criterion 6(6), which provides standards for cleanup of radium; and 10 CFR 51.45(c), which requires the applicant to provide sufficient data in an environmental report for the Commission to conduct an independent analysis.

6.4.5 Reference

NRC. "Multi-Agency Radiation Survey and Site Investigation Manual (MARSSIM)." Revision 1. Washington, DC: NRC. 2000.

6.5   Financial Assurance

6.5.1 Areas of Review

The staff should review financial assessments (cost estimates) provided by the applicant for the costs of ground-water restoration (standard review plan Section 6.1); reclamation (standard review plan Section 6.2); and decommissioning and waste disposal (standard review plan Section 6.3). These assessments may be provided as an appendix. The staff should review provisions for a financial surety that is consistent with Criteria 9 of 10 CFR Part 40, Appendix A, and the guidance in Appendix C of this standard review plan.

6.5.2 Review Procedures

The staff should review the proposed surety amount provided to ensure that it is sufficient to fund all decommissioning activities documented in the license application, that the methods used to establish the surety amount are acceptable, and that the forecast costs are reasonable. Activities to be covered by the surety include reclamation, off-site disposal of 11e.(2) byproduct material, ground-water restoration, structure and equipment removal, and closure. The purpose of the financial surety is to provide sufficient resources for completion of reclamation of the facility including building decommissioning and well field restoration and soil reclamation, by a third party, if necessary.

The reviewer should determine whether the assumptions for the financial surety analysis are consistent with what is known about the site (standard review plan Section 2.0) and the design and operations of the facility and its effluent control system (standard review plan Sections 3.0, 4.0, and 5.0). To the extent possible, the applicant should base these assumptions on experience from generally accepted industry practices, from research and development activities at the site, or from previous operating experience in the case of a license renewal. The values used in the analysis should be based on current dollars (or adjusted for inflation) and reasonable values for the costs of various activities. The reviewer should also examine the type of financial instrument(s) proposed for the surety to ensure that it is consistent with the requirements of 10 CFR Part 40, Appendix A, Criterion 9.

For license renewals and amendment applications, Appendix A to this standard review plan provides guidance for examining facility operations and the approach that should be used in evaluating amendments and renewal applications.

6.5.3 Acceptance Criteria

The cost estimate for ground-water restoration, decommissioning, reclamation, and waste disposal is acceptable if it meets the following criteria:

(1)   The bases for establishing a financial surety in 10 CFR Part 40, Appendix A, Criterion 9, are satisfied. The surety for well fields is usually established as they go into production. Once accepted, the surety will be reviewed annually by NRC to assure that sufficient funds would be available for completion of the reclamation plan by a third party. Detailed guidance on reviewing financial assessments for *in situ* leach operations is found in Appendix C of this standard review plan .

The reviewer shall examine licensee commitments and proposed schedules for surety updates in response to facility changes, annual updates, and changes in closure or decommissioning plans. Additional guidance to reviewers is contained in NMSS Decommissioning Standard Review Plan, NUREG–1727 (NRC, 2000).

(2)   All activities included in the cost estimate are activities that are included either in the reclamation plan or in the operations review completed using Sections 6.1 through 6.4 of this standard review plan.

(3)   All activities included either in the reclamation plan or in Sections 6.1 through 6.4 of this standard review plan are included in the financial analysis.

(4)   The assumptions used for the proposed surety are consistent with what is known about the site (standard review plan Section 2.0) and the design and operations of the facility and its effluent control system (standard review plan Sections 3.0, 4.0, and 5.0). To the extent possible, the applicant has based these assumptions on experience from

generally accepted industry practices, any research and development at the site, or previous operating experience in the case of a license renewal.

(5)   Surety values are based on current dollars (or are adjusted for inflation), and reasonable costs for the required reclamation activities are defined.

(6)   The applicant commits to funding the approved financial surety through one of the mechanisms described in 10 CFR Part 40, Appendix A, Criterion 9, including a (i) surety bond; (ii) cash deposit; (iii) certificate of deposit; (iv) deposit of a government security; (v) irrevocable letters or lines of credit, or (vi) combinations of the above that meet the total surety requirements.

(7)   The applicant commits to updating the surety value annually, in response to changes in closure or decommissioning plans, and as necessitated by changes in the facility and its operations. The annual update will be submitted ninety (90) days prior to the surety anniversary date each year.

(8)   The applicant commits to extending the surety for an additional year if NRC has not approved a proposed revision thirty (30) days prior to the surety expiration date.

(9)   The applicant commits to revising the surety arrangement within three (3) months of NRC approval of a revised closure (decommissioning) plan if estimated costs exceed the amount of the existing financial surety. This revised surety instrument will take effect within thirty (30) days of NRC written approval of the surety documents.

(10)   Surety documentation includes a breakdown of costs; the basis for cost estimates with adjustments for inflation; a minimum 15-percent contingency; and changes in engineering plans, activities performed, and any other conditions affecting estimated costs for site closure.

(11)   The licensee commits to submitting for NRC approval an updated surety to cover any planned expansion or operational change not included in the annual surety update at least ninety (90) days prior to beginning associated construction.

(12)   The licensee commits to providing NRC with copies of surety-related correspondence submitted to a state, a copy of the state's surety review, and the final approved surety arrangement. The licensee also commits that, where the surety is authorized to be held by the state, the surety covers all appropriate costs

(13)   Reclamation/decommissioning plan cost estimates, and annual updates should follow the outline in Appendix C to this standard review plan.

6.5.4 Evaluation Findings

If the staff review, as described in this section, results in the acceptance of the financial assurance cost estimate, the following conclusions may be presented in the technical evaluation report.

NRC has completed its review of the financial assurance cost estimate for the _____ *in situ* leach facility. This review included an evaluation of the methods that will be used to develop the procedures using the review procedures in standard review plan Section 6.5.2 and the acceptance criteria outlined in standard review plan Section 6.5.3.

The applicant has established an acceptable financial assurance cost estimate based on the requirements in 10 CFR Part 40, Appendix A, Criterion 9. The applicant has assured that sufficient funds would be available for completion of the reclamation plan by an independent contractor. The applicant has included in the financial analyses all the activities in the reclamation plan or in Sections 6.1–6.4 of the standard review plan. The applicant has based the assumptions for financial surety analysis on site conditions, including experiences with generally accepted industry practices, research and development at the site, and previous operating experience (in the case of a license renewal). The values used in the financial surety analysis are based on current dollars (or are adjusted for inflation) and reasonable costs for the required reclamation activities are defined. The financial instrument(s) proposed are acceptable to NRC and meet the total surety requirements (select appropriate description).

Based on the information provided in the application and the detailed review conducted of the financial assurance cost estimate for the _____ *in situ* leach facility, the staff concludes that the amount of the proposed financial surety and its methods of estimation are acceptable and are consistent with 10 CFR Part 40, Appendix A, Criterion 9, which requires that financial surety arrangements be established by each operator.

6.5.5 Reference

NRC. NUREG–1727, NMSS Decommissioning Standard Review Plan." Washington, DC: NRC. 2000.

7.0 ENVIRONMENTAL EFFECTS

7.1   Site Preparation and Construction

7.1.1   Areas of Review

The staff should review how construction activities may disturb the existing terrain and wildlife habitats, including the effects of such activities as building temporary or permanent roads, bridges, or service lines; disposing of trash; excavating; and land filling. The staff should also review information on how much land will be disturbed and for how long and whether there will be dust or smoke problems. The staff should review data indicating the proximity of human populations and identifying undesirable impacts on their environment arising from noise; disruption of stock grazing patterns; and inconvenience from the movement of men, material, or machines, including activities associated with any provision of housing, transportation, and educational facilities for workers and their families. Descriptions of any expected changes in accessibility to historic and archeological sites in the region should be assessed. Discussions of measures designed to mitigate or reverse undesirable effects such as erosion control, dust stabilization, landscape restoration, control of truck traffic, and restoration of affected habitats should be reviewed. The staff should also evaluate the beneficial effects of site preparation construction activities, if applicable.

The staff should review the impact of site preparation and construction activities on area water sources and the effects of these activities on fish and wildlife resources, water quality, water supply, aesthetics, as applicable. Reviewers should evaluate measures such as pollution control and other procedures for habitat improvement to mitigate undesirable effects. Staff should consult NUREG–1748 (NRC, 2001) for general procedures for environmental reviews and the environmental assessment process.

The staff should review the resources and ecosystem components cumulatively affected by the proposed action and other past, present, and reasonably foreseeable future actions. The reviewer should examine cumulative impacts by considering whether:

(1)   A given resource is especially vulnerable to incremental effects.

(2)   The proposed action is one of several similar actions in the same geographic area.

(3)   Other activities in the area have similar effects on the resource.

(4)   Effects have been historically significant for this resource.

(5)   Other analyses in the area have identified a cumulative effects concern.

7.1.2   Review Procedures

The staff should determine if the application adequately addresses how site preparation and construction activities may disturb the existing terrain, wildlife habitats, and area water sources in compliance with National Environmental Policy Act Requirements in 10 CFR 51.45 and

Environmental Effects

51.60. The consequences of these activities to both human and wildlife populations should be considered. The descriptions should be adequately supported by site-specific data, well-documented calculations, and accepted modeling studies, as appropriate. The discussion should include those impacts that are unavoidable as well as those that are irreversible. The staff should ensure that the applicant provides information pertaining to how much land will be disturbed and for how long. The staff should confirm that the effects of the following activities and circumstances, where applicable, are addressed: the building of temporary or permanent roads, bridges, or service lines; disposing of trash; excavating and land filling; and the likelihood of dust and smoke problems. The proximity of site activities to nearby human populations should be addressed, as well as anticipated impacts on their environment including noise; disruption of grazing patterns; inconvenience from movement of material and machines; effects arising from additional housing, transportation, and educational facilities for workers and families; and any disruption in access to historic or archeological sites. The staff should ensure that mitigation measures that are adequate to alleviate or significantly reduce environmental impacts are discussed. Examples of mitigation measures include erosion control, dust stabilization, landscape restoration, control of truck traffic, and restoration of affected habitats.

The staff should consider the adequacy of the cumulative impact analysis with respect to past, present, and reasonably foreseeable actions. The staff should determine if the cumulative analysis adequately considered whether and to what extent the environment has been degraded, whether ongoing activities in the area are causing impacts, and trends for activities and impacts in the area. The Council on Environmental Quality has developed guidance (Council on Environmental Quality, 1997) on considering cumulative impacts in the context of National Environmental Policy Act requirements.

The staff should also evaluate any discussion of likely beneficial effects from site preparation and construction to the extent that such might counteract detrimental effects.

For license renewals and amendment applications, Appendix A to this standard review plan provides guidance for examining facility operations and the approach that should be used in evaluating amendments and renewal applications.

7.1.3 Acceptance Criteria

The applicant's assessment of the environmental impacts of site preparation and construction is acceptable if it meets the following criteria:

(1)   All environmental impacts from construction activities are adequately described and supported with site-specific data and, where applicable, modeling studies and calculations.

       A thorough discussion of all construction activities is provided with associated impacts including the generation and control of wastes; dusts; smoke; noise; traffic congestion; disruption of local public services, routines, and property; and aesthetic impacts.

(2)   The applicant adequately describes all unavoidable and irreversible impacts to both the natural environment and nearby human populations.

(3)   The applicant adequately describes the amount of land to be disturbed and the length of time it will be disturbed.

(4)   The applicant has provided an adequate evaluation of the environmental resources that are vulnerable to the incremented effects from the cumulative impacts of the proposed action and other past, present, and reasonably foreseeable action.

(5)   The applicant recommends reasonable mitigation measures for all significant adverse impacts.

(6)   The applicant demonstrates that land can be restored.

7.1.4   Evaluation Findings

If the staff review, as described in this section, results in the acceptance of the environmental assessment of the site preparation and construction plans, the following conclusions may be presented in the technical evaluation report.

NRC has completed its review of the plans for site preparation and construction proposed for use at the _____ in situ leach facility. This review included an evaluation of the methods that will be used to conduct the site preparation and construction using the review procedures in standard review plan Section 7.1.2 and the acceptance criteria outlined in standard review plan Section 7.1.3.

The applicant has acceptably identified all environmental impacts from construction activities including waste generation; dusts; smoke; noise; traffic congestion; disruption of public services, routines, and property; and aesthetic impacts. Applicant plans are supported with site-specific data and modeling studies or calculations, where applicable. Identification and assessment of the effects of all unavoidable and irreversible impacts on the natural environment and humans are acceptable. Disturbance of land and the length and nature of the disturbance are acceptably described. The applicant has recommended appropriate mitigation measures for all significant adverse impacts. The applicant has determined that the land can be returned to its original use after cessation of in situ leach operations.

Based on the Information provided in the application and the detailed review conducted of the site preparation and construction plans for the _____ in situ leach facility, the staff concludes that the environmental impacts of the proposed site preparation and construction are acceptable and are in compliance with 10 CFR 40.32(c), which requires the applicant's proposed equipment, facilities, and procedures be adequate to protect health and minimize danger to life or property; 10 CFR 40.32(d), which requires that the issuance of the license will not be inimical to the common defense and security nor to the health and safety of the public; 10 CFR 40.41(c), which requires the applicant to confine source or byproduct material to the location and purposes authorized in the license; and 10 CFR 51.45(c), which requires the applicant to provide sufficient data for the Commission to conduct an independent environmental analysis.

Environmental Effects

7.1.5 References

Council on Environmental Quality. "Considering Cumulative Effects Under the National Environmental Policy Act." Washington, DC: Council on Environmental Quality, Executive Office of the President. 1997.

NRC. NUREG–1748, "Environmental Review Guidance for Licensing Actions Associated with NMSS Programs." Washington, DC: NRC. 2001.

7.2 Effects of Operations

7.2.1 Areas of Review

The staff should review discussions in the application that address the impact of facility operations on the environment, including surface-water bodies, ground water, air, land, land use, ecological systems, and important plants and animals, as discussed in Section 2.0 of this standard review plan. Staff should consult NUREG–1748 (NRC, 2001) for general procedures for environmental reviews and the environmental assessment process.

7.2.2 Review Procedures

The staff should determine whether the application addresses the impacts of facility operations on the environment, including surface-water bodies, ground water, air, land, land use, ecological systems, and important plants and animals. The staff should determine whether the supporting evidence is based on, and supported by, theoretical, laboratory, onsite, or field studies undertaken for this, or for previous operations.

The staff should determine whether the proposed facility provides for the protection of ground water from the environmental effects of operations. In conducting the review, the staff should focus on (i) characteristics of the hydrological system; (ii) effluent control systems; (iii) spill detection and containment systems in the processing facilities and storage areas; (iv) ground-water monitoring and surface-water monitoring programs, and (v) the ground-water restoration program provided in the application. This information should provide a strong basis for determining the likely overall effects of any impacts to the ground-water system, such as lixiviant excursions, infiltration from spills, or ruptures of wells.

The staff should ensure that, if surface water exists onsite or is connected to off-site surface-water systems, the likely consequences of impacts of operations on surface water are assessed, and mitigation measures are provided. Likely consequences of impacts might include siltation from disruption of surface ground cover or changes to surface drainage patterns. The staff should also determine whether the applicant has assessed the likelihood for decreased air quality resulting from dust loading from truck traffic on dirt roads and exposure of disturbed surface soils to wind. Radiological impacts to air from operations are assessed in other sections of this standard review plan.

In conducting the review, the staff should consider the applicant's ecological information as reviewed in Section 2.8 of this standard review plan to determine if any endangered or sensitive species of plants and animals exist on site. The level of concern for ecological impacts of operations will be affected by the presence of any such sensitive or endangered species. For most facilities, the ecological impacts are expected to be minimal during this period because of the lack of surface disruption during operations. The staff review should ensure that measures have been taken to restrict terrestrial animals from entering facility grounds by use of fencing and other means. In areas used by migrating waterfowl, additional measures may need to be taken to ensure that any surface impoundments are not used by waterfowl. Local ecological conditions may be such that the facility grounds provide favorable habitat for local wildlife, and efforts to minimize contact between wildlife and contaminated areas should be considered. These efforts will serve to mitigate immediate impacts on local species, but will also serve to limit introduction of contamination into the food chain.

For license renewals and amendment applications, Appendix A to this standard review plan provides guidance for examining facility operations and the approach that should be used in evaluating amendments and renewal applications.

7.2.3 Acceptance Criteria

The environmental impacts from operations are acceptable if they meet the following criteria:

(1)   All anticipated significant environmental impacts from facility operations are identified and the applicant provides: (i) mitigation measures for these impacts; (ii) justification for why impacts cannot be mitigated; or (iii) justification for why it is not necessary to mitigate these impacts to protect the local environment.

(2)   At a minimum, the applicant demonstrates that the anticipated impacts on terrestrial and aquatic ecology, air quality, surface- and ground-water systems, land, and land use are environmentally acceptable.

7.2.4 Evaluation Findings

If the staff review, as described in this section, results in the acceptance of the environmental effects of operations, the following conclusions may be presented in the technical evaluation report.

NRC has completed its review of the effects of operations proposed at the _____ in situ leach facility. This review included an evaluation of the effects of operations using the review procedures in standard review plan Section 7.2.2 and the acceptance criteria outlined in standard review plan Section 7.2.3.

The applicant has acceptably described all anticipated significant environmental impacts from facility operations. The applicant has provided acceptable (i) plans to mitigate such impacts; (ii) justification of why impacts cannot be mitigated; or (iii) justification of why it is not necessary to mitigate the impacts to protect the local environment. The applicant has demonstrated that

Environmental Effects

anticipated impacts to terrestrial ecology, air quality, surface- and ground-water systems, and land use are environmentally acceptable.

Based on the information provided in the application and the detailed review conducted of the effects of operations on the _____ in situ leach facility, the staff concludes that the anticipated effects of operations are acceptable and are in compliance with 10 CFR 40.41(c), which requires the applicant to confine source or byproduct material to the location and purposes authorized in the license; and 10 CFR 51.45(c), which requires the applicant to provide sufficient data for the Commission to conduct an independent analysis.

7.2.5 Reference

NRC. NUREG–1748, "Environmental Review Guidance for Licensing Actions Associated with NMSS Programs." Washington, DC: NRC. 2001.

7.3 Radiological Effects

7.3.1 Exposure Pathways

The staff should review information on the radiological effects of operations on humans, including estimates of the radiological impacts from all exposure pathways. The staff should evaluate descriptions of the plant operations with special attention to the likely pathways for radiation exposure of humans. The staff should review information on accumulation of radioactive material in specific internal compartments and should ensure that both internal and external doses are included in the analysis. This information can be tabulated using the outline provided in Appendix A of the Standard Format and Content Guide (NRC, 1982).

7.3.1.1 Exposures from Water Pathways

7.3.1.1.1 Areas of Review

The staff should review the estimates of annual average concentrations of radioactive nuclides in receiving water at the site boundary and at locations where water is consumed or is otherwise used by humans or where it is inhabited by biota of significance to human food chains. The review should include the data presented in support of these estimates, including details of models and assumptions used in supporting calculations of total annual whole body and organ doses to individuals in the off-site population from all receiving water exposure pathways as well as any dilution factors used in these calculations. Additionally, the staff should review estimates of radionuclide concentration in aquatic and terrestrial food chains and associated bioaccumulation factors. The staff should evaluate calculations of internal and external doses. If there are no waterborne effluents from the facility, then these analyses are not needed. Details of models and assumptions used in calculations may be provided in an appendix to the application.

7.3.1.1.2 Review Procedures

The staff should determine whether the concentration estimates at the site boundary meet the regulatory requirements in 10 CFR 20.1302(b)(2)(i) which specifies limits for annual average concentrations of radionuclides in liquid effluents. The staff should also check to ensure that calculations of concentrations have been done for receiving water at locations where water is consumed or is otherwise used by humans or where it is inhabited by biota of significance to human food chains, to meet public dose limits in 10 CFR 20.1301. If the liquid effluent dose is calculated separately from the air pathway dose, the staff should ensure that the results can be summed with the air pathway dose for the total dose comparison to the limit in 10 CFR 20.1301. The staff should also determine whether these estimates are supported by properly interpreted data, calculations, and model results using reasonable assumptions. The staff should review the parameter selections including the justifications provided for important parameters used in the dose calculation. The staff should check the input data for modeling results, to ensure the parameters discussed in the application are the same as those used in the modeling. Code outputs should be spot-checked to ensure that the results are correctly reported in the application. For simple hand calculations, spot calculations can be used to verify that they were done correctly.

For license renewals and amendment applications, Appendix A to this standard review plan provides guidance for examining facility operations and the approach that should be used in evaluating amendments and renewal applications.

7.3.1.1.3 Acceptance Criteria

The exposures from water pathways are acceptable if they meet the following criteria:

(1)   The estimates of individual exposure to radionuclides at the site boundary meet the regulatory requirements in 10 CFR 20.1302(b)(2)(i), which specify limits for annual average concentrations of radioactive nuclides in liquid effluents, or the dose limit in 10 CFR 20.1301.

(2)   Calculations of concentrations of radionuclides in receiving water at locations where water is consumed or is otherwise used by humans or where it is inhabited by biota of significance to human food chains are included in the compliance demonstration for public dose limits in 10 CFR 20.1301.

(3)   For facilities that generate liquid effluents, the relevant exposure pathways are included in a pathway diagram provided by the applicant.

(4)   The conceptual model (scenarios and exposure pathways) is similar to and consistent with the methodology for liquid effluent exposure pathways in Regulatory Guide 1.109, "Calculation of Annual Doses to Man From Routine Releases of Reactor Effluents for the Purpose of Evaluating Compliance With 10 CFR Part 50," Appendix I (NRC, 1977).

(5)   The conceptual model used for calculating the source term and individual exposures (and/or concentrations of radionuclides) from liquid effluents at the facility boundary is

representative of conditions described at the site, as reviewed in Section 2.0 of this standard review plan.

(6)   The parameters used to estimate the source term, environmental concentrations, and exposures are applicable to conditions at the site, as reviewed in Section 2.0 of this standard review plan.

7.3.1.1.4 Evaluation Findings

If the staff review, as described in this section, results in the acceptance of the exposure estimates from water pathways, the following conclusions may be presented in the technical evaluation report.

NRC has completed its review of the radiological effects of exposure from water pathways at the _____ in situ leach facility. This review included an evaluation of the methods that will be used to evaluate radiological effects using the review procedures in standard review plan Section 7.3.1.1.2 and the acceptance criteria outlined in standard review plan Section 7.3.1.1.3.

Applicant estimates of individual exposure to radionuclides from water pathways at the site boundary are acceptable since they are less than the requirements in 10 CFR 20.1302 (b)(2)(i) with regard to annual average concentrations in liquid effluents, or they are less than the dose limit in 10 CFR 20.1301. The applicant has demonstrated that the concentrations of radionuclides in receiving water where it is consumed or otherwise used by humans, or where it is inhabited by biota significant to the human food chain are in compliance with the public dose limits in 10 CFR 20.1301. The applicant has included the relevant pathway diagrams in the application. The applicant has used an acceptable representation of the conditions at the site in the determination of the source term for the model calculations. The applicant has acceptable values for parameters used to estimate the source term, environmental concentrations, and exposures, and the parameters are representative of the _____ in situ leach site.

Based on the information provided in the application and the detailed review conducted of exposures from water pathways for the _____ in situ leach facility, the staff concludes that the exposures from water pathways are acceptable and are in compliance with 10 CFR 20.1302(b)(2)(i), which specifies limits for annual average concentrations of radionuclides in liquid effluents and 10 CFR 20.1301, which specifies dose limits for individual members of the public.

7.3.1.1.5 References

NRC. Regulatory Guide 3.46, "Standard Format and Content of License Applications, Including Environmental Reports, for _In Situ_ Uranium Solution Mining." Washington, DC: NRC, Office of Standards Development. 1982.

————. Regulatory Guide 1.109, "Calculation of Annual Doses to Man From Routine Releases of Reactor Effluents for the Purpose of Evaluating Compliance With 10 CFR Part 50, Appendix I." Washington, DC: NRC, Office of Standards Development. 1977.

7.3.1.2   Exposures from Air Pathways

7.3.1.2.1 Areas of Review

The staff should review estimated release rates of airborne radioactivity from facility operations and the atmospheric dispersal of such radioactivity considering applicable meteorological data as reviewed in Section 2.0 of this standard review plan. The staff should then review the estimates of annual total body and organ doses to individuals including (i) at the point of maximum ground level concentration offsite; (ii) at the site boundary in the direction of the prevailing wind; (iii) at the site boundary nearest the emission source; and (iv) at the nearest residence in the direction of the prevailing wind. The applicant can choose to show compliance with a concentration limit or with individual dose limits. Therefore, the staff should initially determine the method of compliance chosen by the applicant and focus the review accordingly. Regardless of which compliance method is chosen, the reviewer should also evaluate an individual dose to the public to verify compliance with the requirements in 10 CFR 20.1301. The staff should review data, models, calculations, and assumptions used in support of these estimates. The review should consider both the source term and exposure pathway components of the calculation and should include deposition of radioactive material on food crops and pasture grass.

7.3.1.2.2 Review Procedures

The staff should determine whether the estimates of annual total body and organ doses to individuals at the point of maximum ground level concentrations offsite; individuals exposed at the site boundary in the direction of prevailing wind; individuals exposed at the site boundary nearest to the sources of emissions; and individuals exposed at the nearest residence in the direction of the prevailing wind, meet the regulatory requirements in 10 CFR 20.1301. The staff should also determine whether these estimates are supported by properly interpreted data, calculations, and model results using reasonable assumptions.

An acceptable computer code that calculates off-site doses to individuals from airborne emissions from *in situ* leach facilities is MILDOS-AREA (Yuan, et al., 1989). This code does not calculate the source term. Therefore, the applicant must provide documentation of the source term calculation that is used as input to MILDOS-AREA (Yuan, et al., 1989), if this code is used. The staff should review the source term equation to ensure that it is an accurate estimation of all significant airborne releases from the facility including, where applicable, yellowcake dust from the dryer stack and radon emissions from processing tank venting and well field releases. If a closed processing loop is used, then radon release from processing is expected to be negligible. If a vacuum dryer is used for yellowcake, then dust emissions from drying may also be assumed to be negligible. The staff should focus attention on the values used for the production flow and the fraction of this flow that is expected to be released during operations. A reasonable estimate of well field radon release is about 25 percent. The staff should also

Environmental Effects

ensure that the source term calculation accounts for all material released during startup, production, and restoration activities.

The review of the MILDOS-AREA (Yuan, et al., 1989) calculation should focus on the code input provided by the applicant. The applicant should have provided a list of the relevant parameter information that was used. The information from this list should be compared with the input from the code run to ensure that the correct values have been used. Dose results from the code output should be checked against the tabulated results in the application to ensure that the values have been correctly reported. The staff should also evaluate warning messages that the code provides in the output to identify anomalies in the input data or problems with the run. If reported results appear anomalous, the staff may conduct confirmatory analyses using MILDOS-AREA (Yuan, et al., 1989).

For license renewals and amendment applications, Appendix A to this standard review plan provides guidance for examining facility operations and the approach that should be used in evaluating amendments and renewal applications.

7.3.1.2.3 Acceptance Criteria

The exposures from air pathways are acceptable if they meet the following criteria:

(1)   The estimates of individual exposure to radionuclides at the site boundary meet the regulatory requirements in 10 CFR 20.1302(b)(2)(i) with regard to annual average concentrations of radionuclides in airborne effluents or the dose limit in 10 CFR 20.1301. The estimates of individual exposure to radionuclides (not including radon) indicate that the ALARA constraint on air emissions in 10 CFR 20.1101(d) will be met.

(2)   Calculations of concentrations of radionuclides in air at locations downwind where residents live or where biota of significance to human food chains exist are included in the compliance demonstration for public dose limits in 10 CFR 20.1301. The estimates of individual exposures to radionuclides (not including radon) indicate that the as low as is reasonably achievable constraint on air emissions, in 10 CFR 20.1101(d), will be met.

(3)   Relevant airborne exposure pathways are included in the pathway diagram provided by the applicant.

(4)   The conceptual model used for calculating the source term and individual exposures (and/or concentrations of radionuclides) from airborne effluents at the facility boundary is representative of conditions described at the site as reviewed in Section 2.0 of this standard review plan. The conceptual model is consistent with the methodologies described in Regulatory Guide 3.51, Sections 1–3, "Calculational Models for Estimating Radiation Doses to Man From Airborne Radioactive Materials Resulting From Uranium Mill Operations" (NRC, 1982). The conceptual model for the MILDOS-AREA code (Yuan, et al., 1989) is one acceptable method for performing these exposure calculations. Other methods are acceptable if the applicant is able to satisfactorily demonstrate that the model includes the criteria discussed above.

(5)   The parameters used to estimate the source term, environmental concentrations, and exposures are applicable to conditions at the site as reviewed in Section 2.0 of this standard review plan. Guidance on source term calculations is available in Regulatory Guide 3.59, Sections 1–3, "Methods for Estimating Radioactive and Toxic Airborne Source Terms for Uranium Milling Operations" (NRC, 1987). Additionally, an example source term calculation specifically applicable to *in situ* leach facilities is described in Appendix D.

7.3.1.2.4 Evaluation Findings

If the staff review, as described in this section, results in the acceptance of the radiological effects from air pathways, the following conclusions may be presented in the technical evaluation report.

NRC has completed its review of the radiological effects of exposure from air pathways at the _____ *in situ* leach facility. This review included an evaluation of the methods that will be used to evaluate radiological effects using the review procedures in standard review plan Section 7.3.1.2.2 and the acceptance criteria outlined in standard review plan Section 7.3.1.2.3.

Applicant demonstrations of individual exposure to radionuclides from air pathways are acceptable since they are less than the limits in 10 CFR 20.1302 (b)(2)(i) with regard to annual average concentrations in airborne effluents or they are less than the dose limit in 10 CFR 20.1301. The applicant has acceptably demonstrated that the concentrations of radionuclides in air at locations where residents live or where biota of significance to human food chains exist are in compliance with the public dose limits in 10 CFR 20.1301 and the as low as is reasonably achievable constraint on air emissions in 10 CFR 20.1101(d). The applicant has included the relevant airborne exposure pathway diagrams in the application. The applicant has used an acceptable representation of the atmospheric conditions at the site in the determination of the source term and individual exposures for model calculations. The applicant has used acceptable values for parameters used to estimate the source term, environmental concentrations, and exposures; and the parameters are representative of the _____ *in situ* leach site.

Based on the information provided in the application and the detailed review conducted of exposures from air pathways for the _____ *in situ* leach facility, the staff concludes that the exposures from air pathways are acceptable and are in compliance with 10 CFR 20.1302(b)(2)(i), which specifies limits for annual average concentrations of radionuclides in airborne effluents; 10 CFR 20.1301, which specifies dose limits for individual members of the public; and the as low as is reasonably achievable constraint on airborne emissions in 10 CFR 20.1101(d).

7.3.1.2.5 References

NRC. Regulatory Guide 3.59, "Methods for Estimating Radioactive and Toxic Airborne Source Terms for Uranium Milling Operations." Washington, DC: NRC, Office of Standards Development. 1987.

Environmental Effects

———. Regulatory Guide 3.51, "Calculational Models for Estimating Radiation Doses to Man From Airborne Radioactive Materials Resulting From Uranium Milling Operations." Washington, DC: NRC, Office of Standards Development. 1982.

Yuan, Y.C., J.H.C. Wang,, and A. Zielen. "MILDOS-AREA: An Enhanced Version of MILDOS for Large-Area Sources." Report ANL/ES–161. Argonne, Illinois: Argonne National Laboratory, Energy and Environmental Systems Division. 1989.

7.3.1.3 *Exposures from External Radiation*

7.3.1.3.1 Areas of Review

The staff should review estimates of maximum annual external dose that would be received by an individual from direct radiation at the nearest site boundary and in off-site populations. The staff should also review data, models, calculations, and assumptions used in support of these estimates.

7.3.1.3.2 Review Procedures

The staff should determine whether the estimates of maximum annual external dose that would be received by an individual from direct radiation at the nearest site boundary meet the limits specified in 10 CFR 20.1301(a)(2). The staff should also determine whether these estimates are supported by properly interpreted data, calculations, and model results using reasonable assumptions. Staff should confirm that the input parameters used for the external dose calculation are consistent with the information provided in the application. The staff should also confirm that the selected parameter values are representative of conditions at the site as reviewed in Section 2.0 of this standard review plan. Staff should check the source term conceptual model and selected parameter values to ensure that they are appropriate for the site conditions described in the application.

For license renewals and amendment applications, Appendix A to this standard review plan provides guidance for examining facility operations and the approach that should be used in evaluating amendments and renewal applications.

7.3.1.3.3 Acceptance Criteria

The exposures from external radiation are acceptable if they meet the following criteria:

(1)   The estimates of external radiation exposure at the site boundary meet the regulatory limits in 10 CFR 20.1301(a)(2), in accordance with 10 CFR 20.1302(b).

(2)   The applicant provides an exposure pathway diagram that includes the relevant external exposure pathways.

(3)   The model(s) used for calculating the source term, environmental concentrations, and external exposures at the facility boundary are representative of site conditions reviewed in Section 2.0 of this standard review plan.

(4)   The parameters used to estimate the source term, environmental concentrations, and external exposure are applicable to site conditions as reviewed in Section 2.0 of this standard review plan.

7.3.1.3.4 Evaluation Findings

If the staff review, as described in this section, results in the acceptance of the radiological effects of exposures from external radiation, the following conclusions may be presented in the technical evaluation report.

NRC has completed its review of the radiological effects of exposure from external radiation at the _____ *in situ* leach facility. This review included an evaluation of the methods that will be used to evaluate radiological effects using the review procedures in standard review plan Section 7.3.1.3.2 and the acceptance criteria outlined in standard review plan Section 7.3.1.3.3.

Applicant demonstration of individual exposure to radionuclides from external radiation is acceptable and meets the limits in 10 CFR 20.1301(a)(2) in accordance with the requirements of 10 CFR 20.1302 (b). The applicant has provided an acceptable exposure pathway diagram that includes all relevant external pathways. The applicant has used an acceptable representation of the external exposures at the site in the determination of the source term, environmental concentrations, and individual exposures for the model calculations. The applicant has used acceptable values for parameters used to estimate the source term, environmental concentrations, and exposures; and the parameters are representative of the _____ *in situ* leach site.

Based on the information provided in the application and the detailed review conducted of exposures from external radiation for the _____ *in situ* leach facility, the staff concludes that the exposures from external radiation are acceptable and are in compliance with 10 CFR 20.1301(a)(2), which specifies limits for radiation doses in unrestricted areas from external sources in accordance with the methods contained in 10 CFR 20.1302(b).

7.3.1.3.5 References

None.

7.3.1.4   *Total Human Exposures*

7.3.1.4.1 Areas of Review

The staff should review estimates of the maximum annual dose that could be received via all pathways described above by an individual at the site boundary and at the nearest residence. The staff should also review data, models, calculations, and assumptions used in support of these estimates. Much of this review will already have been completed for the pathway-specific calculations, and the total dose will be the sum of these results.

Environmental Effects

7.3.1.4.2 Review Procedures

The staff should determine whether estimates of the maximum annual dose that could be received via all pathways described above by an individual at the site boundary and at the nearest residence meet regulatory requirements in 10 CFR 20.1301. These calculations can be effectively executed by the MILDOS-AREA code (Yuan, et al., 1989). The staff should also determine whether these estimates are supported by properly interpreted data, calculations, and model results using reasonable assumptions. After the pathway-specific calculations have been reviewed, staff should check to ensure that the doses have been correctly summed to determine the total dose. Also, staff should ensure the population dose is compared with a meaningful reference dose, such as that which is expected for the exposure to the same population from background radiation sources.

For license renewals and amendment applications, Appendix A to this standard review plan provides guidance for examining facility operations and the approach that should be used in evaluating amendments and renewal applications.

7.3.1.4.3 Acceptance Criteria

The total human exposure is acceptable if it meets the following criteria:

(1)   The estimates of individual exposure to radionuclides at the site boundary meet the regulatory requirements in 10 CFR 20.1302(b)(2)(i) with regard to annual average concentrations of radioactive nuclides in airborne and liquid effluents or the dose limit in 10 CFR 20.1301.

(2)   Calculations of the maximum individual whole body and organ doses at the site boundary and for the nearest downwind resident and where biota of significance to human food chains exist are included in the compliance demonstration for public dose limits in 10 CFR 20.1301.

(3)   The exposure pathway diagram provided by the applicant includes pathways relevant to all effluents expected from facility operations.

(4)   The models used for calculating the source terms and individual exposures (and/or concentrations of radionuclides) from all effluents at the facility boundary are representative of conditions described at the site as reviewed in Section 2.0 of this standard review plan. The conceptual models are acceptable as described in Sections 7.3.1.1, 7.3.1.2, and 7.3.1.3 of this standard review plan.

(5)   The parameters used to estimate source terms, concentrations, and exposures are representative of conditions described at the site as reviewed in Section 2.0 of this standard review plan.

7.3.1.4.4 Evaluation Findings

If the staff review, as described in this section, results in the acceptance of the radiological effects from total human exposures, the following conclusions may be presented in the technical evaluation report.

NRC has completed its review of the radiological effects of total human exposures at the _____ in situ leach facility. This review included an evaluation of the methods that will be used to evaluate radiological effects using the review procedures in standard review plan Section 7.3.1.4.2 and the acceptance criteria outlined in standard review plan Section 7.3.1.4.3.

Applicant determination of total human exposure to radionuclides at the site boundary is acceptable since it meets the requirements in 10 CFR 20.1301. The applicant has provided an exposure pathway diagram that includes all relevant external pathways. The applicant has used an acceptable representation of the external exposures at the site in the determination of the source term, environmental concentrations, and individual exposures for the model calculations. The applicant has used acceptable values for parameters used to estimate the source term, environmental concentrations, and exposures; and the parameters are representative of the _____ in situ leach site.

Based on the information provided in the application and the detailed review conducted of total human exposures for the _____ in situ leach facility, the staff concludes that the total human exposures are acceptable and are in compliance with 10 CFR 20.1301 which specifies dose limits for individual members of the public.

7.3.1.4.5 Reference

Yuan, Y.C., J.H.C. Wang, and A. Zielen. "MILDOS-AREA: An Enhanced Version of MILDOS for Large-Area Sources." Report ANL/ES–161. Argonne, Illinois: Argonne National Laboratory, Energy and Environmental Systems Division. 1989.

7.3.1.5   Exposures to Flora and Fauna

7.3.1.5.1 Areas of Review

The staff should review estimates of maximum radionuclide concentrations that may be present in important local flora and local and migratory fauna. The staff should also review data, bioaccumulation factors, models, calculations, and assumptions used in support of these estimates.

7.3.1.5.2 Review Procedures

The staff should determine whether estimates of maximum radionuclide concentrations that may be present in important local flora and local and migratory fauna are calculated such that environmental impacts from facility operations can be assessed to address the requirements of 10 CFR Part 51. Particular attention should be paid to impacts to threatened and endangered species. The staff should also determine whether these estimates are supported by properly interpreted data, reasonable bioaccumulation factors, approved calculations, and model results using reasonable assumptions. Detailed biosphere modeling is not necessary for these calculations. Output from MILDOS-AREA (Yuan, et al., 1989) provides ground level concentrations of radionuclides that can then be converted to plant and animal concentrations by use of simple conversion equations that include deposition, uptake factors, plant interception fractions, and animal consumption rates obtained from the literature. The staff should spot-check parameter values against known sources to ensure that they are within expected ranges. The tabulation of bioaccumulation factors and their sources can be presented in an appendix to the application. Provided these concentrations are protective of human health, they would not be expected to adversely affect native plants and animals (Barnthouse, 1995).

For license renewals and amendment applications, Appendix A to this standard review plan provides guidance for examining facility operations and the approach that should be used in evaluating amendments and renewal applications.

7.3.1.5.3 Acceptance Criteria

The exposures to flora and fauna are acceptable if they meet the following criterion:

(1)   The model and parameter values used for calculation of concentrations of radionuclides in important local flora and fauna are consistent with generally accepted health physics practice and are applicable to the species identified at the site, as reviewed in Section 2.0 of this standard review plan.

7.3.1.5.4 Evaluation Findings

If the staff review, as described in this section, results in the acceptance of the radiological effects from exposures to flora and fauna, the following conclusions may be presented in the technical evaluation report.

NRC has completed its review of the radiological effects of exposures to flora and fauna at the _____ in situ leach facility. This review included an evaluation of the methods that will be used to evaluate radiological effects using the review procedures in standard review plan Section 7.3.1.5.2 and the acceptance criteria outlined in standard review plan Section 7.3.1.5.3.

The applicant forecasts that the off-site radiological impacts of operation will be minimal. Flora and fauna in the areas surrounding the project site are similar to those onsite and are common in the region. Since calculated human exposures are protective of human health, they would not be expected to adversely affect the native plants and animals, and as such, are acceptable.

Based on the information provided in the application and the detailed review conducted of exposures to flora and fauna for the _____ in situ leach facility, the staff concludes that the exposures to flora and fauna are acceptable and are in compliance with 10 CFR Part 51 which requires that environmental impacts from facility operations be assessed.

7.3.1.5.5 References

Barnthouse, L.W. "Effects of Ionizing Radiation on Terrestrial Plants and Animals, A Workshop Report." ORNL/TN–13141. Oak Ridge, Tennessee: Oak Ridge National Laboratory. 1995.

Yuan, Y.C., J.H.C. Wang, and A. Zielen. "MILDOS-AREA: An Enhanced Version of MILDOS for Large-Area Sources." Report ANL/ES–161. Argonne, Illinois: Argonne National Laboratory, Energy and Environmental Systems Division. 1989.

7.4   Non-Radiological Effects

7.4.1   Areas of Review

The staff should review estimates of concentrations of nonradioactive constituents in effluents at the points of discharge as compared with natural ambient concentrations and with applicable discharge standards. The review should include the projected effects of the effluents for both acute and chronic exposure of the biota (including any long-term buildup in soils and sediments and in the biota). The staff should evaluate discussions of dilution and mixing of discharge into the receiving environs, and estimates of concentrations at various distances from the point of discharge. The effects on terrestrial and aquatic environments from chemical wastes that contaminate ground water should also be examined.

The staff should also review discussions of any likely consequences of the proposed operation that do not clearly fall under any specific topic previously addressed. These may include changes in land and water use at the project site; sanitary and other recovery plant waste systems; interaction of the facility with other existing or projected neighboring facilities; effects of ground-water withdrawal on ground-water resources in the vicinity of the well field(s) and recovery plant(s); effects of construction and operation of roads, transmission corridors, railroads, et cetera; effects of changes in surface-water availability on biotic populations; and disposal of other solid and liquid wastes.

7.4.2   Review Procedures

The staff should determine whether the estimated concentrations of nonradioactive constituents in effluents at the point of discharge and the projected effects for both acute and chronic exposure of the biota are adequately quantified in accordance with the National Environmental Policy Act requirements in 10 CFR 51.45 and 51.60. Where applicable, the staff should determine whether these estimates are supported by properly interpreted data, reasonable bioaccumulation factors, calculations, and model results using reasonable assumptions.

For license renewals and amendment applications, Appendix A to this standard review plan provides guidance for examining facility operations and the approach that should be used in evaluating amendments and renewal applications.

Environmental Effects

7.4.3 Acceptance Criteria

The non-radiological effects are acceptable if they meet the following criteria:

(1)   The estimated concentrations of nonradioactive wastes in effluents at the point of discharge and the projected effects for both acute and chronic exposure of the biota are adequately quantified in accordance with the National Environmental Policy Act of 1969 requirements in 10 CFR 51.45 and 51.60.

7.4.4 Evaluation Findings

If the staff review, as described in this section, results in the acceptance of the nonradiological effects, the following conclusions may be presented in the environmental assessment.

NRC has completed its review of the nonradiological effects at the _____ in situ leach facility. This review included an evaluation of the methods that will be used to evaluate nonradiological effects using the review procedures in standard review plan Section 7.4.2 and the acceptance criteria outlined in standard review plan Section 7.4.3.

The applicant has acceptably described anticipated significant nonradiological environmental impacts from facility operations. The estimated effects of nonradioactive wastes in effluents at the point of discharge and the projected effects for both acute and chronic exposure of biota are acceptable.

Based on the information provided in the application and the detailed review conducted of nonradiological effects for the _____ in situ leach facility, the staff concludes that the nonradiological effects are acceptable and are in compliance with 10 CFR Part 51.45 which specifies the content of environmental reports.

7.4.5 References

None.

7.5   Effects of Accidents

7.5.1 Areas of Review

The NRC has evaluated the effects of accidents at in situ leach facilities [NUREG–0706 (NRC, 1980); Center for Nuclear Waste Regulatory Analyses, 2001]. These analyses demonstrate that, for most credible potential accidents, consequences are minor so long as effective emergency procedures and properly trained personnel are used. Specific areas where NRC (1980) and Center for Nuclear Waste Regulatory Analyses (2001) indicated that consequences could be significant are (i) radon releases from process streams; (ii) yellowcake dryer explosions; (iii) lixiviant leaks in buried piping between the well fields and the processing facility; and (iv) chemical accidents.

Applicants whose facilities are consistent with the operating assumptions, site features, and designs examined in these NRC analyses need not conduct independent accident analyses. For these applicants, the staff review should focus on accident response procedures and personnel training in their use. Personnel training is evaluated using Section 5.5 of this standard review plan. If an applicant's operating assumptions, site features, and designs are not consistent with these analyses, the applicant must conduct independent accident analyses. In that case, the staff review should evaluate the adequacy of these independent analyses. The scope of this review includes radiological, nonradiological, and transportation accidents. This review should verify that the accident analyses address a spectrum of accidents ranging in severity from trivial to significant, including a characterization of the occurrence rate or probability and likely consequences.

For all applicants, the reviewers should examine standard operating and accident procedures and the training programs for ensuring that personnel can execute them properly. *In situ* leach facility training programs are reviewed using Section 5.5 of this standard review plan.

7.5.2   Review Procedures

For applications that contain independent accident analyses, the staff should determine whether accident scenarios described in the application are reasonable based on descriptions of the facility and operations reviewed in Sections 3.0, 4.0, and 5.0 of this standard review plan and are sufficiently complete to determine environmental impacts of operations pursuant to National Environmental Policy Act requirements. The staff should determine whether these scenarios and estimates are supported by properly interpreted data, calculations, and model results using reasonable assumptions. If consequences cannot be quantified, a qualitative description of the impacts should be reviewed for adequacy. The staff should confirm that uranium extraction industry experience is used to support any accident analyses, including consideration of plant design and specific components that are prone to failure or are known to have failed at other facilities.

For independent analyses of transportation accidents, the staff need not review all operational aspects of transportation activities, as these will be addressed through inspections relevant to the general transportation license requirements.

The staff should ensure the applicant has procedures in place to detect and respond to postulated accident conditions and to mitigate consequences. The reviewers should pay particular attention to procedures related to monitoring, identification, and response to accidents related to: (i) radon release; (ii) yellowcake dryer operations; (iii) leaks in buried lixiviant piping and (iv) chemical releases as they might affect radiological accidents.

For license renewals and amendment applications, Appendix A to this standard review plan provides guidance for examining facility operations and the approach that should be used in evaluating amendments and renewal applications.

Environmental Effects

7.5.3 Acceptance Criteria

The independent analyses of consequences of accidents are acceptable if they meet the following criteria:

(1)   The applicant has provided analyses of credible accident consequences that are consistent with the facility design and planned operations and are sufficient to identify likely environmental impacts from operations.

(2)   Analyses of accident consequences include mitigation measures, as appropriate.

(3)   Analyses of accidents include results from operating experience at similar facilities.

(4)   For radiological accidents, the applicant's response program provides for notification to NRC in compliance with the requirements of 10 CFR 20.2202 and 20.2203.

Adequate procedures to respond to and mitigate or remediate the likely consequences of accidents are identified or referenced in the application.

7.5.4 Evaluation Findings

If the staff's review, as described in this section, results in acceptance of the applicant's description of the effects of accidents, the following conclusions may be presented in the technical evaluation report.

NRC has completed its review of the applicant's description of the effects of accidents for the _____ in situ leach facility. This review included an evaluation of the methods that will be used to evaluate the effects of accidents using the review procedures in standard review plan Section 7.5.2 and the acceptance criteria outlined in standard review plan Section 7.5.3.

The applicant has acceptably described all likely significant effects of accidents from facility operations. The applicant has provided an acceptable analysis of probable accidents and their consequences, if necessary, consistent with facility design, site features, and planned operations. If appropriate, the applicant has confirmed that facility design, site features, and planned operations are consistent with previous NRC accident analyses. The applicant has identified likely environmental impacts from such accidents and has included mitigation measures. Any accident analyses have considered past operating experience from similar facilities. Adequate response and remediation procedures have been identified or referenced, and the facility personnel will be qualified to implement them. The applicant's response program for radiological accidents will comply with the notification requirements of 10 CFR 20.2202 and 20.2203.

Based on the information provided in the application and the detailed review conducted of the effects of accidents for the _____ in situ leach facility, the staff concludes that the effects of accidents are acceptable and are in compliance with 10 CFR Part 51.45, which specifies the content of environmental reports; 10 CFR 40.32(c), which requires that the applicant's proposed equipment, facilities, and procedures be adequate to protect health and minimize danger to life or property; and 10 CFR 20.2202 and 20.2203, which define response program requirements for radiological accidents.

7.5.5 References

Center for Nuclear Waste Regulatory Analyses. NUREG/CR–6733, "A Baseline Risk-Informed, Performance-Based Approach for *In Situ* Leach Uranium Extraction Licenses." San Antonio, Texas: Center for Nuclear Waste Regulatory Analyses. 2001.

NRC. NUREG–0706, "Final Generic Environmental Impact Statement on Uranium Milling—Project M–25." Washington, DC: NRC. September 1980.

7.6 Economic and Social Effects of Construction and Operation

The staff should review descriptions in the application related to the likely economic and social effects of construction and operation of the proposed facility. These impacts should be discussed in separate sections covering benefits, costs, and resources committed.

7.6.1 Benefits

7.6.1.1 Areas of Review

The staff should review social and economic benefits from the proposed *in situ* leach operations that affect various political jurisdictions or public and private interests. Some of these reflect transfer payments or other values that may partially, if not fully, compensate for certain services as well as external or environmental costs, and this fact should be reflected in the designation of the benefit. Some examples of benefits to be reviewed include:

(1)   Tax revenues to be received by local, state, and federal governments.

(2)   Temporary and permanent new jobs created and the associated payroll. (value-added concept)

(3)   Incremental increases in regional productivity of goods and services.

(4)   Enhancement of recreational values.

(5)   Environmental enhancement in support of the propagation or protection of wildlife and the improvement of wildlife habitats.

(6)   Creation and improvement of local roads, waterways, or other transportation facilities.

Environmental Effects

(7)   Increased knowledge of the environment as a consequence of ecological research and environmental monitoring activities associated with plant operation and technological improvements from applicant research programs

The staff should also review discussions of significant benefits that may be realized from construction and operation of the proposed facility, including expressions in monetary terms, discounted to present worth, of who is likely to be affected and for how long. In the case of aesthetic impacts that are difficult to quantify, the staff should review photographs or pictorial drawings of structures or environmental modifications visible to the public.

7.6.1.2   Review Procedures

The staff should determine whether sufficient detail is presented to evaluate significant economic and social benefits that may be realized from construction, operation, restoration, reclamation, and decommissioning of the proposed facility. The staff should determine whether the likely benefits are reasonable and supported by properly interpreted data, calculations, and model results, using reasonable assumptions. The staff should determine to what extent likely benefits can serve to offset adverse effects and costs of construction and operation of the facility. The Standard Format and Contents of License Applications, Including Environmental Reports (NRC, 1982) provides a list of the types of benefits to be included in the application. The NRC has also provided guidance in NUREG–1748 (NRC, 2001) for compliance with requirements of the National Environmental Policy Act.

For license renewals and amendment applications, Appendix A to this standard review plan provides guidance for examining facility operations and the approach that should be used in evaluating amendments and renewal applications.

7.6.1.3   Acceptance Criteria

The economic and social effects of construction and operation are acceptable if they meet the following criteria:

(1)   The applicant's analyses of economic and social benefits that may be realized from construction, operation, restoration, reclamation, and decommissioning of the proposed facility are supported by properly interpreted data, calculations, and model results.

(2)   For each benefit identified, the applicant identifies who is affected and the duration of the impact.

(3)   For special case environmental assessments (e.g., those that have substantial public interest, decommissioning costs involving on-site disposal, decommissioning/ decontamination cases that allow radioactivity in excess of release criteria, or cases where environmental justice issues have been previously raised) the applicant has provided sufficient data to assess environmental justice issues in accordance with NUREG–1748 (NRC, 2001).

7.6.1.4   Evaluation Findings

If the staff review, as described in this section, results in the acceptance of the effects of the economic and social benefits of construction and operation, the following conclusions may be presented in the environmental assessment.

NRC has completed its review of the economic and social benefits of construction and operation proposed at the _____ in situ leach facility. This review included an evaluation of the methods that will be used to evaluate effects of economic and social benefits of construction and operation using the review procedures in standard review plan Section 7.6.1.2 and the acceptance criteria outlined in standard review plan Section 7.6.1.3.

The applicant has acceptably described anticipated economic and social benefits of construction and operation of the facility covering the affected environment and the full extent of activities discussed in Sections 2.0, 3.0, 4.0, 5.0, and 6.0 of the standard review plan. The applicant has provided an acceptable analysis of probable benefits consistent with the facility design and industrywide experience. The applicant has included analyses of: (i) tax revenues; (ii) creation of temporary and permanent jobs and accrued payroll; (iii) incremental increases in regional productivity of goods and services; (iv) enhancement of recreational values; (v) environmental enhancement and increased knowledge of the environment through ecological research and environmental monitoring programs; and (vi) creation and improvement of infrastructure (e.g., roads, waterways, water and power supply, and other transportation facilities). The applicant has acceptably identified for each benefit who is affected and the expected duration of the beneficial effect. Overall, the applicant has demonstrated that the analysis of the economic and social benefits from the construction, operation, restoration, reclamation, and decommissioning of the proposed in situ leach facility are supported by properly interpreted data, calculations, and model results.

Based on the information provided in the application and the detailed review conducted of economic and social benefits of construction and operation for the _____ in situ leach facility, the staff concludes that the economic and social benefits of construction and operation are acceptable and are in compliance with 10 CFR Part 51.45(c) which requires an analysis that balances the impacts of proposed actions.

7.6.1.5   References

NRC. Regulatory Guide 3.46, "Standard Format and Content of License Applications, Including Environmental Reports, for In Situ Uranium Solution Mining." Washington, DC: NRC, Office of Standards Development. 1982.

————. NUREG–1748, "Environmental Review Guidance for Licensing Actions Associated with NMSS Programs." Washington, DC: NRC. 2001.

Environmental Effects

7.6.2 Socioeconomic Costs

7.6.2.1 Areas of Review

The staff should review information presented concerning the ground-water quality restoration, surface reclamation, and plant decommissioning costs; and research and development costs, including postoperational monitoring requirements. The applicant should discount these costs to present worth. Resource commitments are addressed in Section 7.6.3 of this standard review plan.

The staff should also review information on external costs, including the probable number and location of the population group is adversely affected, the estimated economic and social impact, and any special measures taken to alleviate the impact. Environmental justice considerations are presented in NUREG–1748 (NRC, 2001).

Temporary external costs should also be evaluated including housing shortages; inflationary rentals or prices; congestion of local streets and highways; noise and temporary aesthetic disturbances; overloading of utilities, water supply, and sewage treatment facilities; crowding of local schools, hospitals, or other public facilities; overtaxing of community services; and disruption of people's lives or of the local community caused by acquisition of land for the proposed site.

Finally, the staff should review information regarding long-term external costs including: (i) impairment of recreational values (e.g., reduced availability of desired species of wildlife and sport animals, or restrictions on access to land or water areas preferred for recreational use); (ii) deterioration of aesthetic and scenic values; (iii) restrictions on access to areas of scenic, historic, or cultural interest; (iv) degradation of areas having historic, cultural, natural, or archeological value; (v) removal of land from present or contemplated alternate uses; (vi) reduction in quantities of regional products because of displacement of persons from the land proposed for the site; (vii) lost income from recreation or tourism that may be impaired by environmental disturbances; (viii) lost income attributable to environmental degradation; (ix) decrease in real estate values in areas adjacent to the proposed facility; and (x) increased costs to local governments for the services required by the permanently employed workers and their families. In discussing these costs, the applicant should indicate, to the extent practical, who is likely to be affected, to what degree, and for how long.

7.6.2.2 Review Procedures

The staff should determine whether sufficient detail is presented to evaluate significant economic and social internal and external costs that may be incurred during construction, operation, restoration, reclamation, and decommissioning of the proposed facility. The assessment of costs should be reviewed in the context of the information provided in other chapters of the application as reviewed in Sections 2.0, 3.0, 4.0, 5.0, and 6.0 of this standard review plan to ensure consistency and completeness. The staff should review any data, models, calculations, and assumptions used in support of these projections. The staff should ensure the applicant has identified who it is that will bear the cost, the number of such people, the duration of the impacts, and what measures will be taken to mitigate the impacts. Costs should be discounted to present worth. The NRC has provided guidance in NUREG–1748 (NRC, 2001) for compliance with the socioeconomic requirements of the National Environmental Policy Act.

For license renewals and amendment applications, Appendix A to this standard review plan provides guidance for examining facility operations and the approach that should be used in evaluating amendments and renewal applications.

7.6.2.3   Acceptance Criteria

The costs of the *in situ* leach operations are acceptable if they meet the following criteria:

(1)   The analyses of economic and social costs that may be realized from construction, operation, restoration, reclamation, and decommissioning of the proposed facility are supported by properly interpreted data, calculations, and model results.

(2)   For each cost identified, the applicant identifies who is affected, the duration of impacts, and any mitigation measures necessary to alleviate or reduce impacts.

(3)   Costs are discounted to present worth.

7.6.2.4   Evaluation Findings

If the staff review, as described in this section, results in the acceptance of the effects of the economic and social costs of construction and operation, the following conclusions may be presented in the environmental assessment.

NRC has completed its review of the effects of economic and social costs of construction, operation, restoration, reclamation, and decommissioning operations proposed at the _____ *in situ* leach facility. This review included an evaluation of the methods that will be used to evaluate effects of economic and social costs of construction and operation using the review procedures in standard review plan Section 7.6.2.2 and the acceptance criteria outlined in standard review plan Section 7.6.2.3.

The applicant has acceptably described all anticipated economic and social costs of construction and operation of the facility covering the affected environment and the full extent of activities discussed in Sections 2.0, 3.0, 4.0, 5.0, and 6.0 of this standard review plan. The applicant has provided an acceptable analysis of probable costs consistent with the facility design and industrywide experience. The applicant has included analyses of (i) impairment of recreational values; (ii) restriction on access to water or land for recreational use; (iii) restriction on access to areas of scenic, historic, or cultural interest; (iv) deterioration of aesthetic and scenic values; (v) degradation of areas having historic, cultural, natural, or archeological values; (vi) removal of land from present or contemplated alternative uses; (vii) reductions in quantities of regional products; (viii) lost income from recreation or tourism that may be impaired by environmental disturbances; (ix) lost income attributable to environmental degradation; (x) decrease in real estate values adjacent to the proposed facility; and (xi) increased costs to local governments for increased services and infrastructure. The applicant has identified for each cost who is affected, to what extent, and the expected duration of the effect. Overall, the applicant has demonstrated that the analysis of the economic and social costs from the construction, operation, restoration, reclamation, and decommissioning of the proposed *in situ* leach facility is supported by acceptably interpreted data, calculations, and model results.

Based on the information provided in the application and the detailed review conducted of economic and social costs of construction and operation for the _____ *in situ* leach facility, the staff concludes that the economic and social costs of construction and operation are acceptable and are in compliance with 10 CFR Part 51.45(c) which requires an analysis that balances the impacts of proposed actions.

7.6.2.5   *Reference*

NRC. NUREG–1748, "Environmental Review Guidance for Licensing Actions Associated with NMSS Programs." Washington, DC: NRC. 2001.

8.0 ALTERNATIVES TO PROPOSED ACTION

8.1 Areas of Review

The staff will review comparative reconnaissance level evaluations of available alternatives to the licensing action proposed in the *in situ* leach facility application in accordance with the requirements of National Environmental Policy Act of 1969 including realistic alternatives for the various processing stages. As part of this review, the staff should consider the no-action alternative. Alternative designs do not have to be described in as great detail as the proposed action. The purpose of these evaluations is to determine that alternatives that provide a significant reduction in impacts to human health and the environment have not been overlooked. The reviews should include descriptions of the ground-water quality restoration programs to be applied for each alternative other than the no-action alternative. The staff should evaluate alternatives that may reduce or avoid significant adverse environmental, social, and economic effects expected to result from construction and operation of the proposed facility. The staff should also review the bases and rationales for the choices in regard to number, availability, suitability, and factors limiting the range of alternatives that might avoid some or all of the environmental effects identified in Section 7.0 of this standard review plan. The preferred alternative need not be the one with the least adverse impact. For commercial-scale operations, the review should include the comparative evaluation of available alternatives using results obtained from research and development operations, if applicable.

The staff should also review waste management alternatives considering siting, design, and operational performance objectives developed by NRC staff, in addition to the plans for final disposal discussed in Section 6.0 of this standard review plan.

The review should include discussions regarding locating the liquid impoundment areas at sites where disruption and dispersion by natural forces are eliminated or reduced to acceptable levels, and designing the impoundment areas so that seepage of materials into the ground-water system would be eliminated or reduced to acceptable levels.

8.2 Review Procedures

The staff should determine that the applicant has justified the choice of particular economic recovery processes for the mineralized zone by considering and choosing among techniques and processes that affect the environment in minimal ways. The justification should include a comparative evaluation of the available practicable alternatives. Strengths and weaknesses associated with the likely effects of the use of each technique or process, including the ground-water quality restoration program, should be presented. The staff should determine whether the applicant has considered and chosen those alternatives that may reduce or avoid significant adverse environmental, social, and economic effects expected to result from the construction and operation of the proposed facility. The staff should evaluate the basis and rationale the applicant used for the consideration and rating of the alternatives. The staff should determine that, for commercial-scale operations, the comparative evaluation of available alternatives includes results from research and development operations or similar production-scale sites, if appropriate. The preferred alternative need not be the one with the least adverse environmental impact, and the staff shall evaluate whether the proposed action would meet the requirements of 10 CFR Part 40, Appendix A.

Alternatives to Proposed Action

For license renewals and amendment applications, Appendix A to this standard review plan provides guidance for examining facility operations and the approach that should be used in evaluating amendments and renewal applications.

8.3 Acceptance Criteria

The evaluation of alternatives to the proposed action is acceptable if it meets the following criteria:

(1) The applicant considers process alternatives to the proposed action. The applicant identifies alternatives to the operation of the proposed facility in the manner reviewed in Sections 2.0, 3.0, 4.0, 5.0, and 6.0 of this standard review plan that may mitigate adverse environmental, social, and economic effects reviewed in Section 7.0 of this standard review plan. These alternatives may include, but are not limited to:

    (a) The no-action alternative (must be included.)

    (b) Alternative ore extraction processes such as traditional open-pit and underground mining.

    (c) Alternative lixiviant chemistry.

    (d) Alternative ground-water restoration and long-term monitoring techniques.

    (e) Alternative monitoring and waste management practices.

    (f) Uranium recovery process alternatives.

    (g) Construction of a central processing facility versus use of satellite facilities.

(2) The alternatives are compared with the proposed actions considering the site characteristics as reviewed in Section 2.0 of this standard review plan and consistent with existing uranium extraction standards and practices.

The rationale for selecting the proposed method should be provided, and the proposed action should be shown to be at least as effective as the considered alternatives in meeting all regulatory requirements. If the application is for a new commercial-scale license, the consideration should be based on the results of the research and development site, if applicable.

(3) The applicant considers the environmental, social, and economic effects of a no-action alternative. Presumably, the applicant will provide information to demonstrate that the proposed action will provide social and economic benefits that outweigh the environmental impact of operating the facility.

(4)   The applicant clearly identifies the preferred alternative and demonstrates that it would meet the requirements of 10 CFR Part 40, Appendix A.

8.4   Evaluation Findings

If the staff review, as described in this section, results in the acceptance of the alternatives to the proposed action, the following conclusions may be presented in the technical evaluation report.

NRC has completed its review of the alternatives to the proposed action at the _____ *in situ* leach facility. This review included an evaluation of the methods that will be used to develop the alternatives to the proposed action using the review procedures in standard review plan Section 8.2 and the acceptance criteria outlined in standard review plan Section 8.3.

The applicant has considered other alternatives to its proposed *in situ* leach facility such as open-pit or underground mining. Alternatives to the proposed facility operations that might mitigate environmental, social, and economic effects identified in standard review plan Section 7.0 are presented in a form similar to that required in Sections 2.0, 3.0, 4.0, 5.0, and 6.0, of this standard review plan. Alternatives were acceptably considered for lixiviant chemistry, ground-water restoration techniques, waste management practices, and uranium recovery processes. The applicant has demonstrated that the choice of alternative is effective in meeting the applicable requirements of 10 CFR Part 40, Appendix A. Data from past operations or considerations based on results of research and development site were included in the evaluation of the alternatives, as appropriate. The applicant has considered a no-licensing alternative and has demonstrated that the social and economic benefits of the proposed _____ *in situ* leach facility outweigh any adverse environmental impact of the facility.

Based on the information provided in the application and the detailed review conducted of alternatives to the proposed action for the _____ *in situ* leach facility, the staff concludes that the assessment of alternatives to the proposed action is acceptable and is in compliance with 10 CFR Part 51.45(b)(3) which requires that alternatives to the proposed action be analyzed and applicable portions of 10 CFR Part 40, Appendix A, which provides the requirements for extracting source material from ores and for disposal of the associated wastes.

8.5   References

None.

9.0 COST-BENEFIT ANALYSIS

9.1 Areas of Review

The benefit-cost analysis proposed in this section is intended to be a summary of the benefits and costs of the proposed facility. The staff should review the discussion provided and any accompanying illustrations and tables that explain the important benefits and costs of the proposed facility and operations to determine that the issuance of a license is justified. It is important that both quantitative and qualitative justifications be supported with acceptable data and appropriate rationale.

The review should include evaluation against criteria for assessing and comparing benefits and costs where these are expressed in nonmonetary or qualitative terms and rationales for the selection of process alternatives as well as subsystem alternatives. The staff should also evaluate descriptions of any likely cumulative effects, and the rationale for omitting apparent benefits or costs.

The staff should review irreversible and irretrievable commitments of resources caused by the construction, operation, restoration, reclamation, and decommissioning of the proposed facility. This review should include both relative impacts and long-term net effects. Such resources should include permanent land withdrawal, irreversible or irretrievable commitments of mineral resources, water resource needs and ground-water consumption, permanent vegetation and wildlife losses (e.g., unique habitat, species), and consumption of material resources during operation such as processing chemicals and power or energy needs. The staff should review information presented concerning the percentage terms in which the expected resource loss is related to the total resource in the immediate region and in which the immediate region is related to the surrounding regions in terms of affected areas and distances from the site.

9.2 Review Procedures

The reviewer should determine that the benefit-cost statement has been summarized in the form of a narrative and accompanying tables and charts. The important benefits and costs should be contrasted and discussed appropriately to justify the issuance of the license.

The reviewer should determine that the applicant has developed criteria for assessing and comparing benefits and costs where they are expressed in nonmonetary or qualitative terms. Among the criteria that should be considered are (i) ground-water quality or quantity effects; (ii) radiological impact; and (iii) disturbance of the land. The applicant should present the rationales for the selection of process alternatives as well as subsystem alternatives. The reviewer should ascertain that any likely cumulative and symbiotic effects have been detailed along with appropriate rationales for any tradeoffs. If any apparent benefits or costs have been omitted by the applicant, the reviewer should determine that the applicant has presented the rationale for such omissions. The staff should determine that the applicant has related all the terms used in the benefit-cost analysis to the relevant sections of the application. Overall, the benefit-cost section should demonstrate to reviewer satisfaction that the proposed project is a positive economic and social activity.

Cost-Benefit Analysis

The staff should determine whether sufficient detail is presented to evaluate irreversible and irretrievable commitments of resources because of the construction, operation, restoration, reclamation, and decommissioning of the proposed facility. These commitments should be reviewed considering the facility description and operations discussed in other sections of this SRP to ensure consistency and completeness. Resource needs previously identified in existing environmental reports for similar facilities that are currently operating can be used in the staff's review for comparison.

NUREG–1748 (NRC, 2001) provides guidance for compliance with the socioeconomic and cost-benefit considerations required by the National Environmental Protection Act.

For license renewals and amendment applications, Appendix A to this standard review plan provides guidance for examining facility operations and the approach that should be used in evaluating amendments and renewal applications.

9.3   Acceptance Criteria

The benefit-cost analysis is acceptable if it meets the following criteria:

(1)   The economic benefits of the construction and operation of the proposed facility are acceptably summarized. These may include, but are not limited to:

(a)   Tax revenues to be received by federal, state, and local governments.

(b)   Temporary and permanent jobs.

(c)   Incremental increases in regional productivity of goods and service.

(d)   Enhancement of recreational values.

(e)   Environmental enhancement in support of the propagation or protection of wildlife and the improvement of wildlife habitats.

(f)   Creation and improvement of local roads, waterways, or other transportation facilities.

(g)   Increased knowledge of the environment as a consequence of ecological research and environmental monitoring activities associated with plant operation and technological improvements from the applicant's research program.

(2)   Economic benefits are estimated based on realistic assumptions and objective sources such as census data, tax information, and other site characteristics reviewed in Section 2.0 of this standard review plan.

(3)   The applicant provides a summary of the costs of plant decommissioning and site reclamation costs, and ground-water restoration.

(4)   The applicant summarizes short-term external costs as they affect the interests of people other than the owners and operators of the proposed facility. These may include, but are not limited to

  (a)   Housing shortages

  (b)   Local inflation

  (c)   Noise and congestion

  (d)   Overloading of the water supply, water treatment facilities, and disposal landfills

  (e)   Crowding of schools, hospitals, recreational facilities, or other public facilities

  (f)   Disruption of people's lives (e.g., ranching, farming) through the acquisition of land

(5)   The applicant summarizes long-term external costs as they affect the interests of people other than the owners and operators of the proposed facility. These may include, but are not limited to

  (a)   Impairment of recreational values through reduction in wildlife and sport animals

  (b)   Restrictions on access to land or water

  (c)   Aesthetic impacts

  (d)   Degradation or limited access to areas of historical, scenic, or cultural interests

  (e)   Lost income related to limitations on access to land and facilities

  (f)   Decreased real estate values

  (g)   Increased cost to provide government services for increased populations

(6)   The applicant identifies who is most likely to be affected by the construction and operation of the proposed facility, and to the extent possible, identifies how long the disturbance is expected. This information should be consistent with the population information reviewed in Section 2.3 of this standard review plan.

(7)   If the application is for a renewal, the applicant provides a summary of the actual economic benefits and costs of the facility since the last licensing action.

(8)   A comparison of the benefits and costs is presented that acceptably justifies proceeding with the *in situ* leach operations.

(9)   For special case environmental assessments (e.g., those that have substantial public interest, decommissioning cases involving on-site disposal, decommissioning/ decontamination cases that allow radioactivity in excess of release criteria, or cases where environmental justice issues have been previously raised) the applicant has provided sufficient data to assess environmental justice issues in accordance with NUREG–1748 (NRC, 2001).

(10)   The irreversible and irretrievable commitments of resources for the construction, operation, restoration, reclamation, and decommissioning of the proposed facility are appropriate considering the following:

(a)   Permanent land withdrawal

(b)   Permanent commitment of mineral resources

(c)   Permanent commitment of water resources

Post ground-water restoration impacts at public water supply wells are acceptable if the water quality at town wells is consistent with EPA primary and secondary drinking water standards and NRC standards for uranium

(d)   Irreversible loss of surface vegetation

(e)   Irreversible loss of wildlife or wildlife habitat

(f)   Irreversible commitments of material resources including processing chemicals and energy needs

(11)   For each resource area, the applicant identifies who is affected, the duration of impacts, and any mitigation measures proposed as necessary to alleviate or reduce impacts

9.4   Evaluation Findings

If the staff review, as described in this section, results in the acceptance of the cost-benefit analysis, the following conclusions may be presented in the environmental assessment.

NRC has completed its review of the cost-benefit analysis for the _____ in situ leach facility. This review included an evaluation of the methods that will be used to conduct the benefit-cost analysis and the results using the review procedures in standard review plan Section 9.2 and the acceptance criteria outlined in standard review plan Section 9.3.

The applicant has acceptably summarized the social and economic benefits of the construction and operation of the proposed _____ in situ leach facility including (i) additional tax revenues, (ii) temporary and permanent jobs, (iii) incremental increases in regional product, (iv) enhancement of recreational values, (v) environmental enhancement including protection or propagation of wildlife, (vi) creation and improvements in local infrastructure, and (vii) increased awareness of the environment resulting from ecological research and monitoring and any technological improvements resulting from the applicant's program. The applicant has determined economic benefits from objective sources including (i) census data, (ii) tax information, and (iii) other data as evaluated in Section 2.0 of this standard review plan. The applicant has acceptably summarized costs including plant decommissioning, site reclamation, and ground water restoration. The costs for ground-water restoration, decommissioning, and reclamation, as considered in the financial assessment for surety reviewed in Section 6.5 of this standard review plan, are acceptable. The applicant has identified all short-term *in situ* leach facility-driven external costs including (i) housing shortages, (ii) local inflation, (iii) noise and congestion, (iv) overloading of infrastructure (e.g., schools, water supply, transportation links), and (v) disruption of people's lives as a result of land acquisition. The applicant has acceptably determined all facility-driven long-term external costs including (i) impacts on recreation through reduction in wildlife or sport animals; (ii) restrictions to access to land or water; (iii) aesthetic impacts; (iv) degradation or limited access to historic, scenic, or cultural interests; (v) lost income related to limitations on access to land or recreational facilities; (vi) decreased real estate values; and (vii) increased costs to provide government services for any additional population. The applicant has acceptably identified and considered the extent and longevity of the effect of construction and operation on individuals. The applicant has presented a comparison of the societal benefits and costs to society that acceptably justifies the proposed *in situ* leach facility and operations.

The applicant has acceptably described all anticipated economic and social effects of resources committed at the facility covering the affected environment and the full extent of activities discussed in Sections 2.0, 3.0, 4.0, 5.0, and 6.0 of this standard review plan. The applicant has provided an acceptable analysis of probable effects consistent with the facility design and industry-wide experience. The applicant has included analyses of (i) permanent land withdrawal; (ii) permanent commitment of mineral resources; (iii) permanent commitment of water resources; (iv) irreversible loss of surface vegetation; (v) irreversible loss of wildlife or wildlife habitat; and (vi) irreversible commitments of material resources, such as processing chemicals and energy needs. The applicant has acceptably identified, for each resource committed, who is affected, to what extent, and the expected duration of the effect. Overall, the applicant has demonstrated that its analysis of resources committed as a result of the construction, operation, restoration, reclamation, and decommissioning of the proposed *in situ* leach facility is supported by properly interpreted data, calculations, and model results.

Based on the information provided in the application and the detailed review conducted of the benefit-cost analysis for the _____ *in situ* leach facility, the staff concludes that the benefit-cost analysis is acceptable and is in compliance with 10 CFR Part 51.45(c) which requires that economic, technical, and other benefits and costs of the proposed action and alternatives be considered.

9.5   Reference

NRC. NUREG–1748, "Environmental Review Guidance for Licensing Actions Associated with NMSS Programs." Washington, DC: NRC. 2001.

10.0 ENVIRONMENTAL APPROVALS AND CONSULTATIONS

10.1 Areas of Review

The staff should review all licenses, permits, and other approvals of construction and operations required by federal, state, local, and regional authorities for the protection of the environment including a list of those federal and state approvals that have already been received, and the status of those pending approvals. The staff should also review similar information regarding approvals, licenses, and contacts with tribal authorities. The staff should examine previously submitted environmental assessments or environmental impact statements, if appropriate.

The staff should evaluate discussions of the status of efforts to obtain a water quality certification under Section 401 and discharge permits under Section 402 of the Federal Water Pollution Control Act, as amended, if required, including the rationale if certification is not required. The staff should also note the state, local, and regional planning authorities that have been contacted or consulted.

Finally, the staff should review descriptions and records of public meetings and of meetings held with environmental and other citizen's groups with reference to specific instances of the compliance with citizens' group recommendations.

10.2 Review Procedures

The reviewer should determine that the applicant has satisfied all license, permit, and other approvals of construction and operations that are required by federal, state, local, and regional authorities with jurisdiction for the protection of the environment. Types of licenses or permits may include but are not limited to (i) source materials, (ii) underground injection, (iii) surface impoundment construction, (iv) surface discharge, (v) industrial ground-water, (vi) aquifer exemption, (vii) air quality, (viii) disposal well, and (ix) a state *in situ* leach mining permit. The federal and state approvals that have already been received should be listed, and those pending approval should be appropriately identified. The reviewer should determine that the applicant has presented the appropriate environmental assessment or full environmental impact statement for the proposed *in situ* leach site and surrounding area, regardless of whether the assessments are preexisting or prepared especially for this application. This section is intended to cover licensing and permitting of the process as a whole or parts of the process, and does not require a listing of certifications that may be required for equipment or personnel. Copies of associated documentation may be provided as an appendix to the application. NUREG–1748 (NRC, 2001) provides guidance for evaluating compliance with the consultation requirements of the National Environmental Policy Act.

For license renewals and amendment applications, Appendix A to this standard review plan provides guidance for examining facility operations and the approach that should be used in evaluating amendments and renewal applications.

10.3   Acceptance Criteria

The status of environmental approvals and consultations is acceptable if it meets the following criteria:

(1)   The applicant provides a summary of all permits or licenses obtained for the proposed facility. These should clearly identify

   (a)   the type of permit or license

   (b)   The granting authority (local, state, regional, tribal authorities, or federal)

   (c)   The permit or license number (if appropriate)

   (d)   The current status, with expiration date, if appropriate

(2)   For permits not yet granted, the applicant provides a discussion of the current status of the application and objective evidence that the applicant has applied for, but has not yet received, the permit from the granting authority. Such evidence may include copies of documents such as letters from the granting authority or the permit application.

(3)   For permits and licenses not yet granted, the applicant indicates when approval is expected. Consultations with the granting authority can be summarized.

(4)   The granting authority is clearly defined and appropriate to the area being permitted or licensed. If permits are granted under Agreement State status, this should be identified in the application.

(5)   For licenses renewals and amendments, the applicant summarizes public meetings and meetings held with environmental and other citizens' groups since the last licensing application, and responses to the concerns expressed at these meetings.

10.4   Evaluation Findings

If the staff review, as described in this section, results in the acceptance of the environmental approvals and consultations, the following conclusions may be presented in the technical evaluation report.

NRC has completed its review of the environmental approvals and consultations for the _____ in situ leach facility. This review included an evaluation of the methods that will be used to acquire the environmental approvals and consultations using the review procedures in standard review plan Section 10.2 and the acceptance criteria outlined in standard review plan Section 10.3.

The applicant has acceptably identified the environmental approvals and consultations obtained or required for the proposed _____ in situ leach facility. The applicant has provided a summary of all permits and licenses obtained for the proposed facility that identifies the type of permit (license), the granting authority, the assigned number, and the current status with expiration date (if appropriate). For permits not yet received, the applicant has provided a discussion of the status of the application and evidence that the applicant has requested the appropriate permits, and an indication of when the approval is expected. The applicant has identified all permits issued under Agreement State status and demonstrated that the granting authority is appropriate for all permits. Any meetings held with environmental and citizens' groups are acceptably documented.

Based on the information provided in the application and the detailed review conducted of the environmental approvals and consultations for the _____ *in situ* leach facility, the staff concludes that the environmental approvals and consultations are acceptable and are in compliance with 10 CFR 51.45(d) which requires that the environmental report list all federal permits licenses, approvals and other entitlements that must be obtained in connection with the proposed action and describe the status of compliance with these requirements.

10.5 Reference

NRC. NUREG–1748, "Environmental Review Guidance for Licensing Actions Associated with NMSS Programs." Washington, DC: NRC. 2001.

APPENDIX A

GUIDANCE FOR REVIEWING HISTORICAL ASPECTS OF SITE PERFORMANCE FOR LICENSE RENEWALS AND AMENDMENTS

For license renewals and amendments, the historical record of site operations, including air and ground-water quality monitoring, provides valuable information for evaluating the licensing actions. Following are specific areas where a compliance history or record of site operations and changes should be provided for review:

- For license renewals, U.S. Nuclear Regulatory Commission (NRC) inspection reports and license performance reports

- Amendments and changes to operating practices or procedures

- License violations identified during NRC or Agreement State site inspections

- Excursions, incident investigations or root cause analyses, and resultant cleanup histories or status

- Exceedences of any regulatory standard or license condition pertaining to radiation exposure, contamination, or release limits

- Exceedences of any non-radiation contaminant exposure or release limits

- Updates and changes to any site characterization information important to the evaluation of exposure pathways and doses including site location and layout; uses of adjacent lands and waters; population distributions; meteorology; the geologic or hydrologic setting; ecology; background radiological or non-radiological characteristics; and other environmental features

- Environmental effects of site operations including data on radiological and non-radiological effects, accidents, and the economic and social effects of operations

- Updates and changes to factors that may cause reconsideration of alternatives to the proposed action

- For license renewals, updates and changes to the economic costs and benefits for the facility since the last application

- For license renewals, the results and effectiveness of any mitigation proposed and implemented in the original license

If, after a review of these historical aspects of site operations, the staff concludes that the site has been operated so as to protect health and safety and the environment and that no unreviewed safety-related concerns have been identified, then only those changes proposed by the license renewal or amendment application should be reviewed using the appropriate sections of this standard review plan. Aspects of the facility and its operations that have not changed since the last license renewal or amendment should not be reexamined.

APPENDIX B

RELATIONSHIP OF 10 CFR PART 40, APPENDIX A REQUIREMENTS TO STANDARD REVIEW PLAN SECTIONS

The criteria in 10 CFR Part 40, Appendix A were written specifically for conventional uranium recovery facilities. Therefore, they are not all applicable to *in situ* leach facilities. This appendix identifies the specific standard review plan sections where the applicable criteria are addressed.

10 CFR Part 40, Appendix A Criterion	Locations in NUREG–1569 Where the Criterion is Addressed
Criterion 1: Optimize site selection to achieve permanent isolation of tailings without maintenance.	Not applicable.
Criterion 2: Avoid proliferation of small waste disposal sites.	3.1.4, 4.2.4, 6.2.4, 6.3.4
Criterion 3: Dispose of tailings below grade or provide equivalent isolation.	Not applicable.
Criterion 4: Adhere to siting and design criteria.	
(a) Minimize upstream rainfall catchment areas.	Not applicable to *in situ* leach facilities.
(b) Select topographic features that provide good wind protection.	Not applicable to *in situ* leach facilities.
(c) Provide relatively flat embankment and cover slopes.	Not applicable to *in situ* leach facilities.
(d) Establish a self-sustaining vegetative cover or rock cover considering stability, erosion potential, and geomorphology.	Not applicable to *in situ* leach facilities.
(e) Locate away from faults capable of causing impoundment failure.	2.6.4
(f) Design to promote deposition, where feasible.	Not applicable to *in situ* leach facilities.
Criterion 5A: Meet the primary ground-water protection standard.	
(1) Design, construct, and install an impoundment liner that prevents migration of wastes to subsurface soil, groundwater, or surface water.	3.1.4, 4.2.4
(2) Construct liner of suitable materials, place it on an adequate base, and install it to cover surrounding earth likely to be in contact with wastes or leachate.	3.1.4, 4.2.4

10 CFR Part 40, Appendix A Criterion	Locations in NUREG–1569 Where the Criterion is Addressed
(3) Apply alternate design or operating practices that will prevent migration of hazardous constituents into ground water or surface water.	3.1.4, 4.2.4
(4) Design, construct, maintain, and operate impoundments to prevent overtopping.	3.1.4, 4.2.4
(5) Design, construct, and maintain dikes to prevent massive failure.	3.1.4, 4.2.4
Criterion 5B: Conform to the secondary ground-water protection standards.	
(1) Prevent hazardous constituents from exceeding specified concentration limits in the uppermost aquifer beyond the point of compliance.	3.1.4, 5.7.8.4
(2) Define hazardous constituents as those expected to be in or derived from the byproduct material, those detected in the uppermost aquifer, and those listed in Criterion 13.	3.1.4
(3) Exclude hazardous constituents if they are not capable of posing a substantial present or potential hazards to human health or the environment.	3.1.4
(4) Consider identification of underground sources of drinking water and exempted aquifers.	2.2.4, 3.1.4
(5) Ensure hazardous constituents at the point of compliance do not exceed the background concentration, the value in Paragraph 5C, or an approved alternate concentration limit.	3.1.4, 5.7.8.4
(6) Establish alternate concentration limits, if necessary, after considering practical corrective actions, as low as is reasonably achievable requirements, and potential hazard to human health or the environment.	3.1.4
Criterion 5C: Comply with maximum values for ground-water protection.	3.1.4, 5.7.8.4
Criterion 5D: Implement a ground-water corrective action program if secondary ground-water protection standards are exceeded.	5.7.8.4
Criterion 5E: Consider appropriate measures when developing and conducting a ground-water protection program.	

10 CFR Part 40, Appendix A Criterion	Locations in NUREG–1569 Where the Criterion is Addressed
(1) Incorporate leak detection systems for synthetic liners and conduct appropriate testing for clay/soil liners.	4.2.4
(2) Use process designs that maximize solution recycling and water conservation.	4.2.4
(3) Dewater tailings by process devices or properly designed and installed drainage systems.	4.2.4
(4) Neutralize hazardous constituents to promote immobilization.	4.2.4
Criterion 5F: Alleviate seepage impacts where they are occurring and restore ground-water quality.	4.2.4
Criterion 5G: Provide appropriate information for a disposal system.	
(1) Define the chemical and radioactive characteristics of waste solutions.	4.1.4, 4.2.4
(2) Describe the characteristics of the underlying soil and geologic formations.	2.6.4
(3) Define the location, extent, quality, capacity, and current uses of ground water.	2.2.4
Criterion 5H: Minimize penetration of radionuclides into underlying soils when stockpiling.	Not applicable.
Criterion 6: Install an appropriate cover and close the waste disposal area.	
(1) Ensure the cover meets lifetime and radioactive material release specifications.	Not applicable to *in situ* leach facilities.
(2) Demonstrate the effectiveness of the final radon barrier prior to placement of erosion protection barriers or other features.	Not applicable to *in situ* leach facilities.
(3) Demonstrate the effectiveness of phased emplacement of radon barriers as each section is completed.	Not applicable to *in situ* leach facilities.
(4) Document verification of radon barrier effectiveness to the U.S. Nuclear Regulatory Commission (NRC) and maintain records of this verification.	Not applicable to *in situ* leach facilities.
(5) Ensure that radon exhalation is not significantly above background because of the cover material.	Not applicable to *in situ* leach facilities.

10 CFR Part 40, Appendix A Criterion	Locations in NUREG–1569 Where the Criterion is Addressed
(6) Cleanup residual contamination from byproduct material consistent with the radium benchmark dose.	4.2.4, 6.2.4, 6.4.4
(7) Prevent threats to human health and the environment from non-radiological hazards.	2.11.4, 6.2.4
Criterion 6A: Ensure expeditious completion of the final radon barrier. (1) Complete the radon barrier as expeditiously as practical after ceasing operations in accordance with a written, Commission-approved reclamation plan. (2) Extend milestone completion dates if justified by radon release levels, cost considerations consistent with available technology. (3) Authorize disposal of byproduct materials or similar materials from other sources if appropriate criteria are met.	Not applicable.
Criterion 7: Conduct pre-operational and operational monitoring programs.	2.5.4, 5.7.8.4, 5.7.9.4
Criterion 7A: Establish a detection monitoring program to set site-specific ground-water protection standards, a compliance monitoring system once groundwater protection standards have been established, and a corrective action monitoring program in conjunction with a corrective action program.	5.7.8.4, 5.7.9.4
Criterion 8: Conduct milling operations, including ore storage, tailings placement, and yellowcake drying and packaging operations so that airborne releases are as low as is reasonably achievable.	Partially applicable: 4.1.4, 5.3.1.4, 5.3.2.4, 5.7.1.4, 5.7.3.4
Criterion 8A: Conduct and record daily inspections of tailings or waste retention systems and report failures or unusual conditions to NRC.	5.3.1.4, 5.3.2.4
Criterion 9: Establish appropriate financial surety arrangements for decontamination, decommissioning, and reclamation.	6.2.4, 6.5.4
Criterion 10: Establish sufficient funds to cover the costs of long-term surveillance and control.	Not applicable

10 CFR Part 40, Appendix A Criterion	Locations in NUREG–1569 Where the Criterion is Addressed
Criterion 11A: Comply with effectivity dates for site and byproduct material ownership requirements.	Applies to Commission—not addressed in NUREG–1569.
Criterion 11B: Establish license conditions or terms to ensure that licensees comply with ownership requirements prior to license termination for sites used for tailings disposal.	Applies to Commission—not addressed in NUREG–1569.
Criterion 11C: Transfer title to byproduct material and land to the United States or the state in which the land is located.	Not applicable.
Criterion 11D: Permit use of surface and subsurface estates if the public health, safety, welfare, or environment will not be endangered.	Applies to the Commission—not addressed in NUREG–1569.
Criterion 11E: Transfer material and land to the United States or a state without cost other than administrative a legal costs.	Not applicable.
Criterion 11F: Follow specific requirements for land held in trust for or owned by Indian tribes.	Not applicable.
Criterion 12: Minimize or avoid long-term active maintenance and conduct and report on annual inspections.	Applicable to the long-term custodian—not addressed in NUREG–1569.
Criterion 13: Establish standards for constituents reasonably expected to be in or derived from byproduct materials and detected in ground water.	3.1.4

APPENDIX C

RECOMMENDED OUTLINE FOR SITE-SPECIFIC *IN SITU* LEACH FACILITY RECLAMATION AND STABILIZATION COST ESTIMATES

As required under Criterion 9 of 10 CFR Part 40, Appendix A, the licensee shall supply sufficient information for the U.S. Nuclear Regulatory Commission (NRC) to verify that the amount of coverage provided by the financial assurance will permit the completion of all decontamination, decommissioning, and reclamation of sites, structures, and equipment used in conjunction with facility operation. Cost estimates for the following activities (where applicable) should be submitted to NRC with the initial license application or reclamation plan and should be updated annually; as specified in the license. Cost estimates must be calculated on the basis of completion of all activities by a third party (a third party is an independent contractor or operator who is not financially affiliated with the licensee). Unit costs, calculations, references, assumptions, equipment and operator efficiencies, *et cetera*, must be provided. The annual surety estimate must be prospective of all work to be performed at the site. The licensee must provide estimated costs for all decommissioning, reclamation, and ground-water restoration work remaining to be performed at the site, and not simply deduct the cost of work already performed from the previous surety estimate [see NRC Generic Letter 97-03 (NRC, 1997)].

The detailed cost information necessary to verify the cost estimates for the above categories of closure work is summarized in the following recommended outline. For each area, estimates should include costs for equipment; materials; labor and overhead; licenses, permits and miscellaneous site-specific costs; and any other activity or resource that will require expenditure of funds.

(I)   FACILITY DECOMMISSIONING

This includes decommissioning, free release, or disposal of all structures and equipment. This may be accomplished in two phases. In the first phase, only the equipment not used for ground-water restoration (including the stability monitoring period) might be decontaminated, surveyed and released for unrestricted use. Well plugging and removal of the remaining equipment would be performed in a second phase, after the of ground-water restoration has been completed and approved. The buildings used for the *in situ* leach operations may be decontaminated and released for unrestricted use.

(A)   Salvageable building and equipment decontamination (list). For each building or piece of equipment listed, the following cost data should be provided:

(1)   Decontamination

(2)   Refurbishment

(3)   Removal of equipment

(4)   Repairs

Appendix C

(B)   Nonsalvageable building and equipment disposal:

    (1)   List of major categories of buildings and equipment to be disposed of and their corresponding quantities:

        (a)   Structures (list each major) [tons of material and building volume cubic meters (cubic feet)]

        (b)   Foundation concrete [cubic meters (cubic yards)]

        (c)   Process equipment (tons)

        (d)   Piping and insulation (lump sum)

        (e)   Electrical and instrumentation (lump sum)

    (2)   Disposal of chemical solutions within the facility

(C)   Restoration of contaminated areas (process area, affected ground water, surface impoundment residues, etc.)

Removal and Disposal of 11(e).2 byproduct material—Criterion 2 of 10 CFR Part 40, Appendix A, requires that these materials be transported and disposed of at a licensed tailings area or licensed disposal site. The quantity of material to be removed, the distance to the disposal site, and the fees charged by the receiving facility are important considerations in determining the costs of disposal.

Reclamation—This entails recontouring the well fields and surface impoundments and placing top soil or other materials acceptable to the NRC. This may also include revegetation.

    (1)   Removal:

        (a)   Area, depth, and quantity of material to be removed

        (b)   Excavation, loading, transportation, and deposition

    (2)   Revegetation:

        (a)   Area to be revegetated (acre)

        (b)   Obtaining fill material, replacing topsoil, and revegetating

        (c)   Erosion protection

(II) GROUND-WATER RESTORATION AND WELL PLUGGING

In most cases, ground-water restoration consists of ground-water sweeping and water treatment with partial reinjection. The water treatment equipment used during the uranium recovery phase of the operation is generally suitable for the restoration phase. The capital cost of this equipment is usually absorbed during the initial stages of the operation, leaving only the costs of operation, maintenance, and replacement filters for the restoration phase. However, if additional equipment will be required for restoration, associated costs should be detailed here. Replacement costs of some water treatment equipment may need to be included in the surety if the equipment used for restoration is near the end of its serviceable life.

(A)   Method of restoration

(B)   Volume of aquifer required to be restored, area and thickness of aquifer, number of required pumping cycles, and cycling time. The aquifer volume should include the volume of the exploited ore zone, the flow factor, and any contaminated ground water outside the well field (vertical and horizontal excursions)

(C)   Equipment associated with aquifer restoration (e.g., reverse osmosis unit)

(D)   Verification sample analysis

(E)   Well plugging:

    (1)   Number of wells to be plugged

    (2)   Depth and size of each well

    (3)   Material to be used for plugging including acquisition, transportation, and plugging

(III) RADIOLOGICAL SURVEY AND ENVIRONMENTAL MONITORING

Radiological Survey—Surveys and soil samples for radium are required in areas to be released for restricted use. Soils around the well fields, surface impoundments, and process buildings should be analyzed for radium content. A gamma survey of all areas should be made before release for unrestricted use. All equipment released for unrestricted use should be surveyed and the records should be maintained.

(A)   Soil samples

(B)   Decommissioning equipment and building smear samples

(C)   Gamma survey

(D)   Environmental monitoring

Appendix C

(IV)   PROJECT MANAGEMENT COSTS AND MISCELLANEOUS

Itemize estimated costs associated with project management; engineering design, review, and change; mobilization; power during reclamation; quality control; radiological safety; and any other costs not included in other estimation categories.

(V)   LABOR AND EQUIPMENT OVERHEAD, CONTRACTOR PROFIT

Overhead costs for labor and equipment and contractor profit may be calculated as separate items or loaded into hourly rates. If included in hourly rates, the unit costs must identify the percentages applied for each area.

(VI)   CONTINGENCY

The licensee should include a contingency amount to the total cost estimate for the final site closure. The staff considers a 15-percent contingency to be an acceptable minimum amount.

(VIII)   ADJUSTMENTS TO SURETY AMOUNTS

The licensee is required by 10 CFR Part 40, Appendix A, Criterion 9 to adjust cost estimates annually to account for inflation and changes in reclamation plans. The submission should be in the form of a request for amendment to the license.

   (A)   Adjustments for inflation:

   The licensee should submit a revised surety incorporating adjustments to the cost estimates for inflation 90 days before each anniversary of the date on which the first reclamation plan and cost estimate were approved. The adjustment should be made using the inflation rule indicated by the change in the Urban Consumer Price Index published by the U.S. Department of Labor, Bureau of Labor Statistics (http://stats.bls.gov).

   (B)   Changes in Plans:

      (1)   Changes in the process such as size or method of operation

      (2)   Licensee initiated changes in reclamation plans or reclamation/ decommissioning activities performed

      (3)   Adjustments to reclamation plans required by NRC

      (4)   Proposed revisions to reclamation plans with cost estimates and the basis for cost estimates detailed for NRC review and approval.

To avoid unnecessary duplication and expense, NRC shall take into account surety arrangements required by other federal agencies, state agencies, or other local governing bodies. However, the Commission is not required to accept such sureties if they are not sufficient. Similarly, no reduction to surety amounts established with other agencies shall be effected without NRC approval. Copies of all correspondence relating to the surety between the licensee and the state should be provided to NRC. If authorized by NRC to maintain a surety with a state as the beneficiary, it is the responsibility of the licensee to provide NRC with verification of same and ensure that the agreement with the state specifically identifies the financial surety's application, *in situ* leach facility, and decommissioning/ reclamation requirements.

All costs (unit and total) are to be estimated on the basis of third party, independent contractor costs (include overhead and profit in unit costs or as a percentage of the total). Equipment owned by the licensee and the availability of licensee staff should not be considered in the estimate, to reduce cost calculations. All costs should be based on current-year dollars. Credit for salvage value is generally not acceptable in the estimated costs.

NRC staff review may include a comparison of unit cost estimates with standard construction cost guides (e.g., Dodge Guide, Data Quest) and discussions with appropriate state or local authorities (e.g., highway cost construction). The licensee should provide supporting information or the basis for selection of the unit cost figures used in estimates. The staff may elect to use a publicly available computer code such as RACER™ (Talisman Partners, Ltd., 2000) or spreadsheet to assess these costs.

References

NRC. "Annual Financial Surety Update Requirements for Uranium Recovery Licensees." Generic Letter 97-03. Washington, DC: NRC. July 1997.

Talisman Partners, Ltd. "Introduction to RACER 2000™ (Version 2.1.0)—A Quick Reference." Englewood, Colorado: Talisman Partners, Ltd. 2000.

APPENDIX D

MILDOS-AREA: AN UPDATE WITH INCORPORATION OF *IN SITU* LEACH URANIUM RECOVERY TECHNOLOGY

Letter Report

MILDOS-AREA: An Update with Incorporation of *In Situ* Leach Uranium Recovery Technology[*]

E.R. Faillace, D.J. LePoire, S.-Y. Chen, and Y. Yuan[**]

Environmental Assessment Division
Argonne National Laboratory
9700 South Cass Avenue
Argonne, IL 60439

May 1997

[*]Work supported by the U.S. Nuclear Regulatory Commission and the U.S. Department of Energy, Assistant Secretary for Environmental Management, under Contract W-31-109-Eng-38.

[**]Yuan is affiliated with Square Y Consultants, Orchard Park, NY

Appendix D

Contents

Appendix D

Tables

Figures

1.0 Introduction

The MILDOS-AREA computer code was developed at Argonne National Laboratory in 1989 (Yuan, et al, 1989) for evaluating radiological impacts of uranium processing facilities. The code was modified from the original MILDOS code (Strenge and Bander, 1984) to include large-area sources and to incorporate changes in methods for dosimetry calculations. MILDOS-AREA estimates the radiological impacts of airborne emissions of radioisotopes of the uranium-238 series. Two different measures are calculated: dose commitments to human receptors and annual average air concentrations.

MILDOS-AREA incorporated dose conversion factors derived by the International Commission on Radiological Protection (ICRP) recommendations of 1978. The annual average air concentrations were compared with the maximum permissible concentrations (MPCs) in the U.S. Nuclear Regulatory Commission's *Standards for Protection against Radiation* (10 CFR Part 20). On January 1, 1994, a revision to 10 CFR Part 20 (revised Part 20) went into effect. The revised Part 20 updated its dosimetry to the ICRP 1978 recommendations. The dose limit to the general public also changed. The changes led to a revision of the calculated allowable concentrations for unrestricted areas, with MPC being replaced by the term "effluent concentrations." Therefore, the calculations performed by MILDOS–AREA were not consistent with the current terminology and data contained in the revised Part 20.

In addition, a new method of recovering uranium gained popularity in the late 1980s, and now the majority of operating licensees use the *in situ* leach (ISL) method. In a typical ISL mining site (Hunter, 1996), a licensee uses a series of injection wells that introduce dissolved oxygen and sodium carbonate/bicarbonate into the ore zone. The uranium is mobilized and is extracted through a series of pumping wells. The uranium-rich water is routed through a processing building, where the uranium is removed from the water by ion-exchange (IX) columns. The loaded IX resin is then processed to remove the uranium (elution). The eluted uranium is further processed into a concentrated uranium slurry. The slurry is then dried into yellow cake (U_3O_8). The dried U_3O_8 is packaged and shipped for further processing into enriched uranium and reactor fuel.

Some ISL facilities have smaller processing plants remote from the main processing plant. These plants, called satellite facilities, generally will collect the uranium in resin tanks and then ship the loaded resin to the main processing plant for elution, drying, and packaging. The satellite facilities allow the licensee to economically mine uranium a distance away from the main processing plant.

2.0 Project Objectives

The overall objective of this project is to update the MILDOS-AREA code data structures and terminology to be consistent with revised 10 CFR Part 20. Another objective is the creation of an example problem for ISL facilities. Finally, the above objectives result in the creation of a patch program that will update current versions of MILDOS-AREA to the new version.

This report consists of three components: (1) modification of the data structure of the MILDOS-AREA code, (2) source term derivation for the ISL mining technology, and (3) application of this methodology in the sample problem. Finally, a computer patch program containing this updated information is described. This patch program is to be attached to MILDOS-AREA as an update for the particular application.

3.0 Modifications to The Mildos-area Code

Two sets of modifications are made to the MILDOS-AREA code. These changes reflect both the semantic and the dosimetric revisions implemented in the revised 10 CFR Part 20.

The first modification consists of replacing all occurrences of MPC with allowable concentration (ALC). These changes affect the last page(s) of output for each time step, where the concentrations of radionuclides in air at each receptor location are reported. These pages are now referred to as the "Results of the ALC Check at this Location."

The second modification consists of replacing the old MPC values in the MILDOS-AREA database with the numbers currently tabulated under Effluent Concentrations (Air - Column 1) in Table 2 of Appendix B to the revised 10 CFR Part 20. An exception is radon-222 (Rn-222), where the ALC is expressed in units of working level (WL). The value for Rn-222 is derived as specified in the text of Appendix B; to revised Part 20; the occupational derived air concentration of 1/3 WL has been divided by 300. Table 3-1 lists the radionuclides and the ALCs used in MILDOS-AREA.

TABLE 3-1 Allowable Concentrations Used in Mildos-area

Radionuclide	AC (Inhalation Class) (pCi/m^3)	Default Inhalation Class
Uranium-238	3(D), 1(W), 0.06(Y)	Y
Uranium-234	3(D), 1(W), 0.05(Y)	Y
Thorium-230	0.02(W), 0.03(Y)	W
Radium-226	0.9 (W)	W
Radon-222	1/900 (*)	(*)
Lead-210	0.6 (D)	D
Bismuth-210	500 (D), 40 (W)	W
Polonium-210	0.9 (D), 0.9 (W)	W

(*) Radon-222 is gaseous; the AC is reported in Wls.

4.0 Source Term Estimation for a Sample Isl Facility

The sources of radioactive effluent from an operating ISL uranium recovery facility include (1) the drilling operation at new well fields, (2) uranium extraction operations at production well fields, (3) drying and packaging of yellow cake, (4) restoration operations at old well fields, and (5) land application areas. The following sections describe a methodology for source term derivation for ISL sites that may be used instead of the methodology presented in NUREG/CR-4088 (Hartley, et al, 1985). *Other methodologies may be more appropriate for a particular operating site.*

4.1 New Well Field

Conventional rotary rigs are commonly employed for all drilling activities at an ISL facility. Because all exploration drill holes are sealed with high-viscosity bentonitic mud to maintain aquifer isolation, no particulates are expected to be released during drilling operations. The only source of radioactive release is the Rn-222 from radium-containing ore cuttings temporarily stored in the mud pit. During the period when the ore cuttings are awaiting disposal while stored in a mud pit, radioactive decay of radium-226 (Ra-226) is producing radon continuously. The amount of Rn-222 available for release, or the maximum release rate, in a year as a result of Ra-226 decay from ore cuttings in storage is assumed to be given by the following expression:

$$Rn_{nw} = 10^{12} \, E \, L \, [Ra] \, T \, M \, N \qquad \qquad (1)$$

where

Rn_{nw}	=	Rn-226 release rate from new well field (Ci/yr),
10^{12}	=	unit conversion factor (Ci/pCi),
$[Ra]$	=	concentration of Ra-226 in ore (pCi/g),
E	=	emanating power (dimensionless),
L	=	decay constant of Rn-222 (0.181/d),
T	=	storage time in mud pit (d),
M	=	average mass of ore material in the pit (g), and
N	=	number of mud pits generated per year.

4.2 Production Well Field

No particulate materials are expected to be released from the production well field because its process streams, from production and injection wells to IX columns in the satellite facility, are all in a closed-loop circuit. The primary radioactive emission from the process streams of the production well field is Rn-222 gas. In the natural environment, radon emanates continuously in the ground and migrates through the rock or soil by both diffusion and convection. The movement of radon in ground water in most cases is governed by water transport, rather than by diffusion (Hess, et al, 1985; Mueller Associates, Inc., 1986). In an ISL production well field, the radon released from the ore body is readily removed by the process water ("lixiviant") moving through the well field by injection and production wells. The 3.8-day half-life of Rn-222 allows it to circulate along with the process water in the well field over a long time before it decays.

Appendix D

The general equation describing the change in Rn-222 concentration over time in the process water of a well field can be expressed as:

$$V \frac{dC_{Rn}}{dT} = f\,S - (L + v)\,V\,C_{Rn} - (F_p + F_i)\,C_{Rn} \qquad (2)$$

where

V	=	volume of water in circulation (L),
C_{Rn}	=	Rn-222 concentration in process water (pCi/L),
f	=	fraction of radon source carried by circulating water (dimensionless),
S	=	radon source (pCi/d),
L	=	decay constant of Rn-222 (0.181/d),
v	=	rate of radon venting from piping and valves during circulation (1/d),
F_p	=	"purge" rate of treated water (L/d), and
F_i	=	water discharge rate from resin unloading of IX columns (L/d).

The balance of the fraction of radon source carried by circulating water accounts for any radon in the mined area that is not swept into the injection-production well loop and remains trapped in the ore zone. The "purge" or "bleed" in the production well field is necessary to maintain a hydraulic cone of depression around each well field to prevent leakage of mining solutions outside the production zone.

The radon source term, S, can be expressed as

$$S = 10^6 \times L\,E\,[Ra]\,A\,D\,P \qquad (3)$$

where

10^6	=	unit conversion factor (cm^3/m^3),
E	=	emanating power of active ore zone (dimensionless),
$[Ra]$	=	Ra-226 concentration in ore zone (pCi/g),
A	=	active area of ore zone (m^2),
D	=	average thickness of ore zone (m), and
P	=	bulk density of ore material (g/cm^3).

The water discharge rate from resin unloading, F_i, can be calculated by

$$F_i = N_i\,V_i\,P_i \qquad (4)$$

where

V_i	=	volume content of IX column (L),
N_i	=	number of IX column unloadings per day, and
P_i	=	porosity of resin material.

Under steady-state conditions, the Rn-222 concentration in the process water, C_{Rn}, can be written as

$$C_{Rn} = \frac{10^6\,[Ra]\,A\,D\,P\,E\,L\,f}{(L+v)\,V + F_p + F_i} \qquad (5)$$

When pressure is reduced during purging or when water is aerated during irrigation, radon is readily released to the atmosphere. The amount of Rn-222 available for release from the "purge" is dependent on the water volume purge rate, F_p, and on the Rn-222 concentration in the purged liquid, C_{Rn}. By conservatively assuming that all available radon in the purge water is released, the annual Rn-222 emission is

$$Rn_w = 3.65 \times 10^{10}\,C_{Rn}\,F_p \qquad (6)$$

where
    3.65×10^{10} = unit conversion factor (Ci/pCi)(d/yr), and
    Rn_w = Rn-222 release rate from purge water (Ci/yr).

The annual Rn-222 releases from occasional venting from wellheads and leaking transport piping are

$$Rn_v = 3.65 \times 10^{10} \, v \, C_{Rn} \, V \tag{7}$$

where Rn_v is the annual Rn-222 release from venting (Ci/yr).
The annual radon-222 discharge from the unloading of the IX column contents is

$$Rn_x = 3.65 \times 10^{10} \, F_i \, C_{Rn} \tag{8}$$

where Rn_x annual Rn-222 release from unloading of IX column content (Ci/yr).

The total annual Rn-222 release from the production well field is the sum of Rn_w, Rn_v, and Rn_x. The occurrence of radon in water is controlled by the chemical concentration of radium in the host soil or rock and the emissivity of radon into water. Radon enters air-filled pores in the soil mainly because of the recoil of radon atoms on the decay of Ra-226. The fraction of radon formed in the soil which enters the pores is called the emanating power; reported values range from about 1% to 80%, with an average of 20%, depending on soil type, pore space, and water content (Mueller Associates, Inc., 1986). Varying environmental conditions have been found to affect the rate of radon emanation. In particular, moisture has been found to have significant effects on the radon emanation rate. For purposes of conservatively estimating the radon release from ISL well fields, the emanating power is assumed to be 0.25.

4.3 Drying and Packaging of Yellow Cake

For facilities using rotary vacuum dryers for processing yellow cake, no particulate emissions are expected under normal operating conditions. For facilities using thermal drying, stack releases may be

estimated on the basis of information provided by a number of operating ISL uranium recovery facilities. Although more data are needed, the stack release of yellow cake has been estimated to be about 0.05% of the amount produced; however, because the day-to-day variations of particulate release rates can vary by several times, the assumption is that 0.1% of the uranium produced escapes as particulates into the atmosphere, as suggested in the *Final Generic Environmental Impact Statement on Uranium Milling* (U.S. Nuclear Regulatory Commission, 1980).

The particulate release of nuclides other than uranium isotopes is estimated by grab samples reported by ISL facilities (e.g., Semiannual Reports for Highland Uranium Project, Irigary and Christensen Ranch Projects, Crownpoint, and others). On the basis of the field measurements, the conservative assumption is that the activities of thorium (0.15-0.4% of measured values), radium (0.2-0.3%), lead, polonium, and its decay progeny are 0.5% of the U-238 activity in the yellow cake. Furthermore, it may be assumed that the fraction of this activity that is released is the same as the fraction of uranium (0.1%) that is released.

4.4 Restoration Well Field

The basic operating processes of the restoration well field are similar to those of the production well field. Ground water affected by leaching processes in the production well fields is restored to its premining levels (1) by the "pump and treat" (ground-water sweep) method and by flushing with fresh water injection, and (2) by using the permeative stream from reverse-osmosis treatment units. Like the production well field, no particulate materials are expected to be released from the restoration well field operations. The primary source of radioactive release is the Rn-222 gas in the process water circulating within and discharged from the restoration operations. The annual Rn-222 releases from the restoration well field therefore can be calculated by Equations 6 and 7.

4.5 Releases from Land Application Areas

Radionuclide-containing water, either from purge water from production well fields or from restoration wastewater from restoration well fields, is treated to unrestricted release levels and disposed of by irrigation. Release onto the soil surface will contaminate the soil at the land application areas. The radionuclides adsorbed by the soil will become a source term for radioactive release through wind erosion processes. To estimate this wind-generated source term by using MILDOS-AREA, the radionuclide concentration in the soil needs to be estimated first. The radionuclide concentration in the contaminated surface soil region of the land application area, C_s, is calculated by

$$C_s = \frac{10^3\, C_{tw}\, V_o\, R_s}{A_s\, S_d\, P_s} \qquad (9)$$

where

$\quad C_s$ = radionuclide concentration in the surface soil (pCi/g),

$\quad 10^{-3}$ = unit conversion factor (L/cm^3),

$\quad V_o$ = total volume of water released onto the land application area (m^3),

$\quad C_{tw}$ = radionuclide concentration in treated water (pCi/L),

$\quad A_s$ = area of land application (m^2),

$\quad S_d$ = assumed depth of contaminated area (m),

$\quad P_s$ = bulk density of surface soil (g/cm^3), and

$\quad R_s$ = fraction of radionuclide in irrigation water retained in the soil particles (dimensionless).

The fraction of radionuclides in irrigation water retained in the soil particles, R_s, can be calculated with the following formula:

$$R_s = \left(1 - \frac{1}{R_d}\right) \qquad (10)$$

The retardation factor, R_d, can be calculated with the following formula:

$$R_d = 1 + \frac{P_s K_d}{w} \tag{11}$$

where

K_d =   radionuclide distribution coefficient (cm^3/g), and

w =   soil volume water content (dimensionless).

The volumetric water content of the soil, w, is the fraction of the total porosity of the soil material occupied by water. The radionuclide distribution coefficient is the ratio of the radionuclide equilibrium concentration of the adsorbed radionuclide in soil to the desorbed radionuclide in water. Representative distribution coefficients can be found in the report by Yu, et al, 1993.

5.0 Example of Source Term Calculation For Sample Isl Facility

The following example illustrates some typical calculations that may be used to derive the source term at a hypothetical operating ISL uranium recovery facility. The example covers the potential operations that may result in radionuclide releases to the air from a typical facility. Note that reasonable assumptions for input parameters have been used for this hypothetical site, *but these input data are not intended to serve as substitutes for data collected at actual operating facilities.*

The layout of the hypothetical site is shown in Figure 5-1. It consists of a main processing facility, a satellite facility, one well field under development (active well field 1, two production well fields (active well fields 2 and 3), a restoration well field, two radium-settling ponds (P1 and P2), a holding pond, and an irrigation plot. Only small portions of the well fields are assumed to be active over any one-year period of operations. Eight receptor locations are identified. Of these, location 5 is included within a cattle grazing area to estimate the dose from consumption of livestock products that may become contaminated from site releases. Source and receptor locations are reported in kilometers east (x coordinate) and north (y coordinate) of the dryer stack in the main processing facility. Negative values of x and y coordinates indicate west and south directions, respectively. Table 5-1 lists the coordinates, used in the input data file for each source and receptor. The meteorology for the site is assumed to be the generic file provided with the code.

Figure 5-1. Layout of Hypothetical ISL Facility

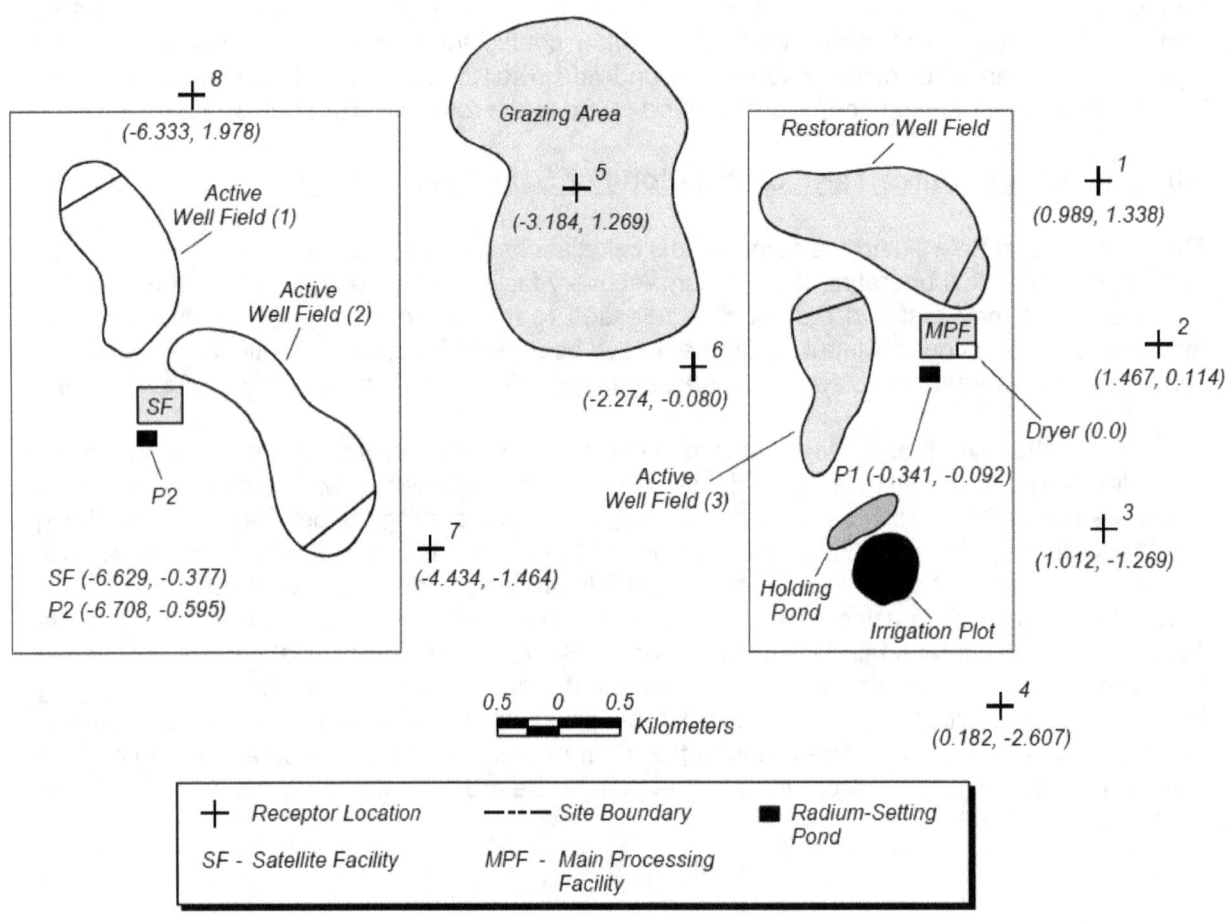

5.1 Summary of Principal Operating Characteristics of the Sample ISL Facility

The following parameters apply to the entire facility:
Yellow cake production rate = 520 metric ton (MT)/yr
Average ore activity, U-238 and each progeny in secular equilibrium = 280 pCi/g
Ore porosity = 0.28
Ore density = 1.8 g/cm^3

5.2 New Well Field Drilling/Construction Area (Well Field 1):

A portion of well field 1, located north of the satellite facility, is under development, as follows:
Number of new wells per peak year = 600
Number of new wells per mud pit = 12
Number of mud pits = 600/12 = 50
Ore zone thickness = 5 m
Drill hole diameter = 8 in.
Average ore material per well (g)=3.14 × (8 in / 2 × 2.54 cm/in)^2 × 500 cm ×
    1.8 g/cm^3 = 2.9 × 10^5
Total ore material in mud pit per year (g) = 3.5 × 10^6
Average storage time of ore grade material in mud pits = 12d
Radon emanating power = 0.25

TABLE 5-1 <u>Source and Receptor Coordinates</u>

Source (km)	East (km)	North (km)	Receptor	East (km)	North
1. Yellow Cake Dryer Stack	0.000	0.000	Receptor 1 (Individual)	0.989	1.338
2. Main Processing Facility IX Columns	0.000	0.000	Receptor 2 (Individual)	1.467	0.114
3. Satellite-Facility	6.629	0.377	Receptor 3 (Individual)	1.012	1.269
4. Radium-Setting Pond 1	0.341	0.092	Receptor 4 (Individual)	0.182	2.607
5. Radium-Setting Pond 2	6.708	0.595	Receptor 5 (Grazing)	3.184	1.269
6. Active Well Field 1 (Area Source)	7.363	1.162	Receptor 6 (Individual)	2.274	0.08
	7.380	1.313	Receptor 7 (Individual)	4.434	1.464
	7.145	1.464	Receptor 8 (Individual)	6.333	1.978
	6.893	1.380			
7. Active Well Field 2 (Area Source)	5.449	1.489			
	4.879	1.053			
	5.080	1.438			
	5.282	1.556			
8. Active Well Field 3 (Area Source)	1.423	0.307			
	1.305	0.525			
	1.104	0.575			
	0.886	0.441			
9. Restoration Well Field (Area Source)	0.248	0.407			

Source (km)	East (km)	North (km)	Receptor	East (km)	North
	0.054	0.927			
	0.137	0.575			
	0.014	0.374			
10. Irrigation Plot (Area Source)	0.669	1.825			
	0.830	1.704			
	0.952	1.448			
	0.911	1.448			

For this location, on the basis of an average Ra-226 concentration of 280 pCi/g, the annual Rn-222 emission from the mud pit can be estimated by using Equation 1:

$$Rn_{nw} = 10^{-12} \text{ Ci/pCi} \times 0.25 \times 0.181/d \times 280 \text{ pCi/g} \times 12 \text{ d} \times 3.5 \times 10^6 \text{ g} \times 50/yr$$
$$= 0.027 \text{ Ci/yr}$$

The radon flux can then be estimated by dividing the total emission rate by the area under development as follows:

Area of active drilling per year = 60,000 m^2

Average Rn-222 flux rate = $(10^{12} \text{ pCi/Ci} \times 0.027 \text{ Ci/yr}) / [60,000 \text{ m}^2 \times (3.15 \times 10^7 \text{ s/yr})]$
= 0.0143 pCi/m^2/s

5.3 Production Well Field 2

The following assumptions are used for the production well field located just to the east of the satellite facility:

Operating days per year = 365
Dimensions of the active ore body:
Peak area per year to be mined = 50,000 m^2
Average thickness of ore bodies = 3 m
Total flow volume in circulation in well field = 50,000 × 3 × 0.28 = 42,000 m^3
= 4.2 × 10^7 L
The following assumptions are made for the satellite facility:
Dimensions or capacity of resin column = 3,500 gal
Resin porosity = 0.4
Number of loaded resin unloadings per day = 3
Water discharge rate from unloading of IX column

= 3,500 gal × 0.4 × 3.785 L/gal × 3/d = 1.6×10^4 L/d
Total wastewater "purge" rate = 100 gallons per minute (gpm)
= 100 gpm × 3.785 L/gal × 60 min/h × 24 h/d = 5.5×10^5 L/d
Fraction of radon source carried by circulating water = 0.8
Rate of radon venting during circulation = 0.01/d
The radon concentration in circulating water is derived by using Equation 5[*] :

C_m = [(10^6 × 280 × 50,000 × 3 × 1.8 × 0.25 × 0.181) × 0.8] /
      {[0.191 × (4.2×10^7)] + [(5.5×10^5) + (1.6×10^4)]}
    = [(3.4×10^{12}) × 0.8] / (8.6×10^6) = 3.2×10^5 pCi/L

The radon release rate from purge water into settling pond P2 is derived by using Equation 6:

Rn_w = (3.65×10^{10})   (3.2×10^5)   (5.5×10^5)
     = 64 Ci/yr

The radon release rate from gas venting and leaking during circulation is derived by using Equation 7:

Rn_v = (3.65×10^{10}) × 0.01 × (3.2×10^5) × (4.2×10^7)
     = 49 Ci/yr

The radon release rate from IX unloading is derived by using Equation 8:

Rn_x = (3.65×10^{10}) × (3.2×10^5) × (1.6×10^4)
     = 1.9 Ci/yr

The total radon release from production well field 2 = 115 Ci/yr.

5.4 Production Well Field 3

The following assumptions are used for the production well field located just to the west of the main processing facility:

Operating days per year = 365
Dimensions of the active ore body:
Peak area per year to be mined = 55,000 m^2
Average thickness of ore bodies = 5 m
Total flow volume in circulation in well field
= 55,000 × 5 × 0.28 = 77,000 m^3 = 7.7×10^7 L

The same parameters used for the satellite facility servicing well field 2 apply to the IX facility used for well field 3. The following source terms have been derived by using Equations 5 to 8.
The radon concentration in circulating water for well field 3 is given by

C_m = [(10^6 × 280 × 55,000 × 5 × 1.8 × 0.25 × 0.181) × 0.8] /
      {[0.191 × (7.7×10^7)] + [(5.5×10^5) + (1.6×10^4)]}
    = [(6.3×10^{12}) × 0.8] / (1.53×10^7) = 3.3×10^5 pCi/L

The radon release rate from purge water into settling pond P1 is given by

Rn_w = (3.65×10^{10}) × (3.3×10^5) × (5.5×10^5)
    = 66 Ci/yr

The radon release rate from gas venting and leaking during circulation is given by

Rn_v = (3.65×10^{10}) × 0.01 × (3.3×10^5) × (7.7×10^7)
    = 93 Ci/yr

---

[*]To reduce the length of this and other calculations, most of the units have been omitted. The reader is referred back to the equations in Chapter 4 for details on parameter descriptions and units.

The radon release rate from IX unloading is given by
$$Rn_x = (3.65 \times 10^{10}) \times (3.3 \times 10^5) \times (1.6 \times 10^4)$$
$$= 1.9 \text{ Ci/yr}$$
The total radon release from production well field 3 = 161 Ci/yr.

5.5 Restoration Well Field

The following assumptions were used for the restoration well field north of the main processing facility:

Expected restoration operation time = 7 yr
Operating days per year = 240
Dimensions of restoration ore body:
Area per year to be restored = 100,000 m^2
Average thickness of ore bodies = 5 m
Total flow volume in circulation in well field
= 100,000 × 5 × 0.28 = 140,000 m^3 = 1.4 × 10^8 l
Total treated water "purge" rate = 200 gpm
= 200 gpm × 3.785 L/gal × 60 min/h × 24 h/d = 1.1×10^6 L/d
Fraction of radon source carried by circulating water = 0.8
Rate of radon venting during circulation = 0.01/d

The following source terms have been derived by using Equations 5 to 7.
The radon concentration in circulating water for the restoration well field is given by

$$C_m = [(10^6 \times 280 \times 100{,}000 \times 5 \times 1.8 \times 0.25 \times 0.181) \times 0.8] /$$
$$\{[0.191 \times (1.4 \times 10^8)] + (1.1m \times 10^6)]\}$$

$$= [(1.1 \times 10^{13}) \times 0.8] / (2.8 \times 10^7) = 3.3 \times 10^5 \text{ pCi/L}$$

The radon release rate from purge water into settling pond P1 is given by

$$Rn_w = (240/365) \times (3.65 \times 10^{-10}) \times (3.3 \times 10^5) \times (1.1 \times 10^6)$$

$$= 87 \text{ Ci/yr}$$

The radon release rate from gas venting and leaking during circulation is given by

$$Rn_v = (240/365) \times (3.65 \times 10^{10}) \times 0.01 \times (3.3 \times 10^5) \times (1.4 \times 10^8)$$

$$= 110 \text{ Ci/yr}$$

The total radon release from the restoration well field = 197 Ci/yr.

5.6 Land Application (Irrigation) Area

The following assumptions are made for the irrigation plot:

Radionuclide concentrations in the holding pond:

U-238 = 1,200 pCi/L
Th-230 = 5 pCi/L

Ra-226 and all progeny = 30 pCi/L
Land irrigation operation water flow rate = 400 gpm
 = 400 gpm × 3.785 L/gal × 60 min/h × 24 h/d = 2.2×10^6 L/d
Land irrigation operation = 122 d/yr
Land irrigation operation lifetime = 7 yr
Total volume water released over operation lifetime
= (2.2×10^6 L/d) × 122 d/yr × 7 yr × 10^{-3} m³/L = 1.9×10^6 m³
Total area of clean wastewater land application = 185,000 m²
Assumed depth of contaminated area = 0.15 m
Density of soil = 1.6 g/cm³
Soil volume water content = 0.25
Distribution coefficient of soil (cm³/g):
Uranium = 50
Thorium = 60,000
Radium = 70
Lead = 100
The retardation factors of surface soil, calculated by using Equation 11, are
Uranium = 320
Thorium = 380,000
Radium = 450
Lead = 640
The fraction of radionuclides in irrigation water that is retained in the surface soil, calculated by using Equation 10, is
Uranium = 1
Thorium = 1
Radium = 1
Lead = 1
The land application area peak surface soil radionuclide concentrations, calculated by using Equation 9, are
U-238 = (10^{-3} × 1,200 × 1.9×10^6 × 1) / (185,000 × 0.15 × 1.6)
        = 0.043   1,200 = 51 pCi/g
Th-230 = 0.043   5 = 0.21 pCi/g
Ra-226 = 0.043   30 = 1.3 pCi/g
Pb-210 = 0.043   30 = 1.3 pCi/g
Radon flux = 1.3 pCi/g   1.0 (pCi/m²/s)/(pCi/g) = 1.3 (pCi/m²/s)

5.7 Main Processing Facility

The following assumptions apply to the main processing facility:
Yellow cake (U_3O_8) production = 520 MT/yr
Stack release rate:
U-238
= 520 MT/yr × 0.001 × 10^6 g/MT × 0.85 g U-nat/g U_3O_8 × (3.3×10^7 Ci U-238/g U-nat)
= 0.146 Ci/yr
Th-230
= 0.146 × 0.005 = 0.00073 Ci/yr
Ra-226, Pb-210, and Po-210
= 0.146 × 0.005 = 0.00073 Ci/yr

6.0 DESCRIPTION OF PATCH PROGRAM

The revisions to the MILDOS-AREA code are incorporated in the following files:

MILMAIN.EXE. This file is the FORTRAN executable file containing the revisions discussed in Chapter 3. It replaces the old MILMAIN.EXE.

SAMPISL.DAT. This file is the input data file for the example ISL facility described in Chapter 5. A copy of the input data file and output file can be obtained upon request to the U.S. Nuclear Regulatry Commission.

MILDOS.UPD. This data file contains the updated allowable concentration levels for the radionuclides listed in Table 3-1.

README.TXT. This text file contains instructions to MILDOS-AREA on how to replace the old MILMAIN.EXE with the new version and how to copy the other two files to the user's MILDOS directory.

7.0 REFERENCES

Hartley, J.N. et al., 1985, *Methods for Estimating Radioactive and Toxic Airborne Source Terms for Uranium Milling Operations*, NUREG/CR-4088, PNL-5338, prepared by Pacific Northwest Laboratory, Richland, Wash., for the U.S. Nuclear Regulatory Commission, Washington, D.C., June.

Hess, C.T., et al., 1985, "The Occurrence of Radioactivity in Public Water Supplies in the United States," *Health Physics, vol.* 48, No. 5, May.

Hunter, J., 1996, "Making a Success of In-Situ Leaching at the Highland Uranium Project," presented at the annual meeting of the Society for Mining, Metallurgy, and Exploration, Phoenix, Ariz., March 11-14.

Mueller Associates, Inc., 1986, *Indoor Air Quality Environmental Information Handbook: Radon*, DOE/PE/720132-2, prepared by Mueller Associates, Inc., Baltimore, Md., for U.S. Department of Energy, Washington, D.C.

Strenge, D.L., and T.J. Bander, 1984, *MILDOS — A Computer Program for Calculating Environmental Radiation Doses from Uranium Recovery Operations*, NUREG/CR-2011, PNL-3767, Pacific Northwest Laboratory, Richland, Wash., for the U.S. Nuclear Regulatory Commission, Washington, D.C.

U.S. Nuclear Regulatory Commission, 1980, *Final Generic Environmental Impact Statement on Uranium Milling*, NUREG-0760, Washington, D.C., Sept.

Yu, C., et al., 1993, *Manual for Implementing Residual Radioactive Material Guidelines Using RESRAD, Version 5.0*, ANL/EAD/LD-2, Argonne National Laboratory, Argonne, Ill., Sept.

Yuan, Y.C., J.H.C. Wang, and A.J. Zielen, 1989, *MILDOS-AREA: An Enhanced Version of MILDOS for Large-Area Sources*, ANL/ES-161, Argonne National Laboratory, Argonne, Ill., June.

APPENDIX E

GUIDANCE TO THE U.S. NUCLEAR REGULATORY COMMISSION STAFF ON THE RADIUM BENCHMARK DOSE APPROACH

E1.0 Background

In 10 CFR 40.4, byproduct material is defined as the tailings or waste produced by the extraction or concentration of uranium or thorium from any ore processed primarily for its source material content, including discrete surface wastes resulting from uranium solution extraction processes. Uranium milling is defined as any activity resulting in byproduct material. Therefore, 10 CFR Part 40, Appendix A, applies to *in situ* leach, heap leach, and ion-exchange facilities that produce byproduct material, as well as to conventional uranium and thorium recovery facilities. This guidance only addresses uranium recovery facilities because there are no currently licensed or planned thorium recovery facilities.

The final rule, "Radiological Criteria for License Termination of Uranium Recovery Facilities," became effective on June 11, 1999, and added the following paragraph after the "radium in soil" criteria in Appendix A, Criterion 6(6):

Byproduct material containing concentrations of radionuclides other than radium in soil, and surface activity on remaining structures, must not result in a total effective dose equivalent exceeding the dose from cleanup of radium contaminated soil to the above standard (benchmark dose), and must be at levels which are as low as is reasonably achievable. If more than one residual radionuclide is present in the same 100-square-meter area, the sum of the ratios for each radionuclide, of concentration present to the concentration limit, will not exceed 1 (unity). A calculation of the peak potential annual total effective dose equivalent within 1,000 years to the average member of the critical group that would result from applying the radium standard (not including radon) on the site, must be submitted for approval. The use of decommissioning plans with benchmark doses which exceed 100 mrem/yr, before application of as low as is reasonably achievable, requires the approval of the Commission after consideration of the recommendation of the staff. This requirement for dose criteria does not apply to sites that have decommissioning plans for soil and structures approved before June 11, 1999.

E2.0 Radium Benchmark Dose Approach

The general requirements for a decommissioning plan, including verification of soil contamination cleanup, are addressed in Chapter 6.0 of the standard review plan. This appendix discusses the NRC staff evaluation of the radium benchmark dose approach, specifically dose modeling and its application to site cleanup activities that should be addressed in the decommissioning plan for those uranium recovery facilities licensed by the NRC and subject to the new requirements for cleanup of contaminated soil and buildings under 10 CFR Part 40, Appendix A, Criterion 6(6), as amended in 1999. The facilities that did not have an approved decommissioning plan at the time the rule became final are required to reduce residual radioactivity, that is, byproduct material, as defined by 10 CFR Part 40, to levels based on the potential dose, excluding radon, resulting from the application of the radium (Ra-226) standard at the site. This is referred to as the radium benchmark dose approach.

Appendix E

This guidance also applies to any revised decommissioning plan submitted for NRC review and approval, after the final rule is effective. However, if a subject licensee can demonstrate that no contaminated buildings will remain, and that soil thorium-230 (Th-230) does not exceed 5 pCi/g (above background) in the surface and 15 pCi/g in subsurface soil in any 100-square-meter area that meets the radium standard, and the natural uranium (U-nat, i.e., U-238, U-234, and U-235) level is less than 5 pCi/g above background, radium benchmark dose modeling is not required. If future modeling with site-specific parameters for uranium recovery sites indicates that this is not a protective approach, the guidance will be revised. Therefore, it would be prudent for a uranium recovery licensee to consider the potential dose from any residual thorium and uranium.

The unity "rule" mentioned in the new paragraph of Criterion 6(6) applies to all licensed residual radionuclides. Therefore, if the ore (processed by the facility), tailings, or process fluid analyses indicate that elevated levels of Th-232 could exist in certain areas after cleanup for Ra-226, some verification samples in those areas should be analyzed for Th-232 or Ra-228. The thorium (Th-232) chain radionuclides (above local background levels) in milling waste would have soil cleanup criteria similar to the uranium chain radionuclides. The staff considers the EPA memorandum of February 12, 1998, (Directive No. 9200.4–25) concerning use of 40 CFR Part 192 soil criteria for Comprehensive Environmental Response, Compensation and Liability Act sites, an acceptable approach. This means that the Th-230 and Th-232 should be limited to the same concentration as their radium progeny with the 5 pCi/g (0.19 Bq/g) criterion applying to the sum of the radium (Ra-226 plus Ra-228) as well as the sum of the thorium (Th-230 plus Th-232) above background.

E2.1   Radium Benchmark Dose Modeling

E2.1.1 Areas of Review

The radium benchmark dose approach involves calculation of the peak potential dose for the site resulting from the 5 pCi/g [0.19 Bq/g] concentration of radium in the surface 15 cm [6 in.] of soil. The dose from the 15 pCi/g [0.56 Bq/g] subsurface radium would also be calculated for any area where the criterion is applied. The dose modeling review involves examining of the computer code or other calculations employed for the dose estimates, the code or calculation input values and assumptions, and the modeling results (data presentation).

Evaluation of the radium benchmark dose modeling as proposed in the decommissioning plan, requires an understanding of the site conditions and site operations. The relevant site information presented in the plan or portions of previously submitted documents (e.g., environmental reports, license renewal applications, reclamation plan, and characterization survey report) should be reviewed.

E2.1.2 Review Procedures

The radium benchmark dose modeling review consists of ascertaining that an acceptable dose modeling computer code or other type of calculation has been used, that input parameter values appropriate (reasonable considering long-term conditions and representative of the application) for the site have been used in the modeling, that a realistic (overly conservative is not acceptable as it would result in higher allowable levels of uranium or thorium which would not be as low as is reasonably achievable) dose estimate is provided, and that the data presentation is clear and complete.

E2.1.3 Acceptance Criteria

The radium benchmark dose modeling results will be acceptable if the dose assessment (modeling) meets the following criteria:

(1)   Dose Modeling Codes and Calculations

The assumptions are considered reasonable for the site analysis, and the calculations employed are adequate. Reference to documentation concerning the code or calculations is provided [e.g., the RESRAD Handbook and Manual (Argonne, 1993a,b)].

The RESRAD code developed by the U.S. Department of Energy (Version 6.1, 2001) may be acceptable for dose calculations because, although the RESRAD ground-water calculations have limitations, this does not affect the uranium recovery sites that have deep aquifers (ground-water exposure pathway is insignificant). The DandD code developed by the NRC (see website ftp://nwerftp.nwer.sandia.gov/nrc/DandD/; also see http://techconf.llnl.gov/radcri/ then dose assessment) provides conservative default values, but does not, at this time, allow for modeling subsurface soil contamination and does not allow calculation of source removal due to soil erosion. Neither the RESRAD nor the DandD code would be adequate to model the dose from off-site contamination, but codes such as GENII are acceptable. See Appendix C of NUREG–1727 (NRC, 2000) for additional information.

If the code or calculations assumptions are not compatible with site conditions, adjustments have been made in the input to adequately reflect site conditions. For example, the RESRAD code assumes a circular contaminated zone. The shape factor (external gamma, code screen R017) must be adjusted for an area that is not circular.

The code and/or calculation provides an estimated annual dose as total effective dose equivalent in mrem/yr. The DandD code provides the annual dose, but RESRAD calculates the highest instantaneous dose. However, RESRAD results are acceptable for long-lived radionuclides that do not move rapidly out of surface soils.

(2)   Input Parameter Values

The code/calculation input data are appropriate for the site and represent current or long-term conditions, whichever is more applicable to the time of maximum dose. When code default values are used, they are justified as appropriate (representative) for the site. Excessive conservatism (i.e., upper bound value) is not used, as this would result in a higher dose and thus higher levels of uranium and thorium could be allowed to remain on site.

Appendix E

Previously approved MILDOS code input parameter values may not be appropriate, because derived operational doses in the restricted area may be an order of magnitude higher than acceptable doses for areas to be released for unrestricted use.

Site-specific input values are demonstrated to be average values of an adequate sample size. Confidence limits are provided for important parameters so that the level of uncertainty can be estimated for that input value. Alteration of input values considers that some values are interrelated [see draft NUREG–1549, Appendix C (NRC, 1998)], and relevant parameters are modified accordingly. The preponderance of important parameter values are based on site measurements and not on conservative estimates. One or more models consider the annual average range of parameter values likely to occur within the next 200 years, for important parameters that can reasonably be estimated. Some other considerations for the input parameter values follow:

(a)   Scenarios for the Critical Group and Exposure Pathways

The scenario(s) chosen to model the potential dose to the average member of the critical group[1] from residual radionuclides at the site reflect reasonable probable future land use. The licensee has considered ranching, mining, home-based business, light industry, and residential farmer scenarios, and has justified the scenarios modeled.

On the basis of one or more of these projected (within 200 years is reasonably foreseeable) land uses to define the critical group(s), the licensee has determined and justified what exposure pathways are probable for potential exposure of the critical group to residual radionuclides at the site. Dairies are not likely to be established in the area of former uranium recovery facilities because the climate and soil restrict feed production. Even if some dairy cows were to graze in contaminated areas, the milk would probably be sent for processing (thus diluted), and not be consumed directly at the site. Therefore, milk consumption is not a likely ingestion exposure pathway. Also, a pond in the contaminated area providing a significant quantity of fish for the resident's diet is not likely, so the aquatic exposure pathway may not have to be modeled. However, the external gamma, plant ingestion, and inhalation pathways are likely to be important.

The radon pathway is excluded from the benchmark dose calculation as defined in Criterion 6(6) of Appendix A to 10 CFR Part 40. This also reflects the approach in the decommissioning rule (radiological criteria for license termination, 10 CFR Part 20, Subpart E).

---

[1]As defined in 10 CFR Part 20, "the group of individuals reasonably expected to receive the greatest exposure to residual radioactivity for any applicable set of circumstances."

(b)   Source Term

If the RESRAD code is used, the input includes lead-210 (Pb-210) at the same input value as for Ra-226. The other radium progeny are automatically included in the code calculations. The chemical form of the contamination in the environment is considered in determining input values related to transport, or inhalation class (retention in the lung) for dose conversion factors.

(c)   Time Periods

The time periods for calculation of the dose from soil Ra-226 include the 1,000-year time frame. The calculated maximum annual dose and the year of occurrence are presented in the results.

(d)   Cover and Contaminated Zone

A cover depth of zero is used in the surface contamination model, and a depth of at least 15 cm [6 in.] is used for the subsurface model. The values for area and depth of contamination are derived from site characterization data. The erosion rate value for the contaminated zone is less than the RESRAD default value because in regions drier than normal, the erosion rate is less, as discussed in the RESRAD Data Collection Handbook (Argonne, 1993a), and the proposed value is justified. The soil properties are based on site data (sandy loam or sandy silty loam are typical for uranium recovery sites), and other input parameters are based on this demonstration of site soil type [see RESRAD handbook, pp., 23, 29, 77, and 105 (Argonne, 1993a)].

The evapotranspiration coefficient for the semi-arid uranium recovery sites is between 0.6 and 0.99. The precipitation value is based on annual values averaged over at least 20 years, obtained from the site or from a nearby meteorological station.

The irrigation rate value may be zero, or less than a code's default value, if supported by data on county or regional irrigation practices (e.g., zero is acceptable if irrigation water is obtained from a river not a well). The runoff coefficient value is based on the site's soil type, expected land use, and regional morphology.

(e)   Saturated Zone

The dry bulk density, porosity, "b" parameter, and hydraulic conductivity values are based on local soil properties. The hydraulic gradient for an unconfined aquifer is approximately the slope of the water table. For a confined aquifer, it represents the difference in potentiometric surfaces over a unit distance.

If the RESRAD code is used, the non-dispersion model parameter is chosen for areas greater than 1,000 square meters (code screen R014), and the well pump rate is based on irrigation, stock, or drinking water well pump rates in the area.

(f)   Uncontaminated and Unsaturated Strata

The thickness value represents the typical distance from the soil contamination to the saturated zone. Since the upper aquifer at uranium recovery sites is often of poor quality and quantity, the depth of the most shallow well used for irrigation or stock water in the region is chosen for the unsaturated zone thickness. A value of 18 m [60 ft] is typical for most sites {15 m [50 ft] for the Nebraska site}, but regional data are provided for justification. The density, porosity, and "b" parameter values are similar to those for the saturated zone, or any changes are justified.

(g)   Distribution Coefficients and Leach Rates

The distribution coefficient (Kd) is based on the physical and chemical characteristics of the soil at the site. The leach rate value of zero in the RESRAD code is acceptable as it allows calculation of the value. If a value greater than zero is given, the value is justified.

(h)   Inhalation

An average inhalation rate value of approximately 8,395 m^3/yr is used for the activity assumed for the rancher or farmer scenario based on a draft letter report (Sandia, 1998a). The mass loading for inhalation (air dust loading factor) value is justified based on the average level of airborne dust in the local region for similar activities as assumed in the model.

(i)   External Gamma

The shielding factor for gamma is in the range of 0.4 to 0.8 (60 to 20 percent shielding) based on DandD Parameter data (NRC, 1998) (the DandD code screening default value is 0.55). The factor is influenced by the type (foundation, materials) of structures likely to be built on the site and the gamma energy of the radionuclides under consideration.

The time fractions for indoor and outdoor occupancy are similar to default values in RESRAD and draft guidance developed for the decommissioning rule [NUREG/CR–5512, Volume 3 (NRC, 1996)]. For example, the staff would consider fraction values approximating 0.7 indoors and 0.15 outdoors for a resident working at home, and 0.5 outdoors and 0.25 indoors for the farmer scenario (the remaining fraction allocated to time spent off site).

The site-specific windspeed value is based on adequate site data. The average annual windspeed for the uranium recovery sites varies from 3.1 to

5.5 meters/sec [7 to 13 mph]. The maximum and annual average windspeed are also considered when evaluating proposed erosion rates.

(j)   Ingestion

Average consumption values (g/yr) for the various types of foods are based on average values as discussed in NUREG/CR–5512, Volume 3 (NRC, 1996), or the Sandia Draft Letter Reports (1998a,b), or are otherwise justified. Livestock ingestion parameters are default values, or are otherwise justified.

For sites with more than 100 acres of contamination, the fraction of diet from the contaminated area is assumed to be 0.25 for the farmer scenario (Sandia, 1998a), or is otherwise justified based on current or anticipated regional consumption practices for home-grown food. Because of the low level of precipitation in the areas in which uranium recovery facilities are located, extensive gardens or dense animal grazing is not likely, so the percentage of the diet obtained from contaminated areas would be lower than the code default value.

Note that the default plant mass loading factor in the DandD code can reasonably be reduced to 1 percent (Sandia, 1998c). The depth of roots is an important input parameter for uranium recovery licensees using the RESRAD code. The value is justified based on the type of crops likely to be grown on the site in the future. For vegetable gardens, a value of 0.3 is more appropriate than the RESRAD default value of 0.9 meters that is reasonable for alfalfa or for a similar deep-rooted plant.

(3)   Presentation of Modeling Results

The radium benchmark dose modeling section of the decommissioning plan includes the code or calculation results as the maximum annual dose (total effective dose equivalent) in mrem/yr, the year that this dose would occur, and the major exposure pathways by percentage of total dose. The modeling section also includes discussion of the likelihood of the various land-use scenarios modeled (reflecting the probable critical groups), and provides the variations in dose (dose distribution) created by changing key parameter values to reflect the range of dose values that are likely to occur on the site. The section also contains the results of a sensitivity analysis (RESRAD can provide a sensitivity analysis via the graphics function) to identify the important parameters for each scenario.

E2.1.4 Evaluation Findings

If the staff review, as described in this section, results in the acceptance of the radium benchmark dose modeling, the following conclusions may be presented in the technical evaluation report.

Appendix E

The staff has completed its review of the site benchmark dose modeling for the
_____ uranium *in situ* leach facility. This review included an evaluation using the review procedures and the acceptance criteria outlined in Section 2.1 of Appendix E of the *in situ* leach standard review plan.

The licensee has provided an acceptable radium benchmark dose model, and the staff evaluation determines that: (1) the computer code or set of calculations used to model the benchmark dose is appropriate for the site; (2) input parameter values used in each dose assessment model are site-specific or reasonable; and (3) the dose modeling results include adequate estimates of dose uncertainty.

On the basis of the information presented in the application, and the detailed review conducted of radium benchmark dose modeling for the _____ uranium *in situ* leach facility, the staff concludes that the information is acceptable and is in compliance with 10 CFR Part 40, Appendix A, Criterion 6(6), which provides requirements for soil and structure cleanup.

E2.2   Implementation of the Benchmark Dose

E2.2.1 Areas of Review

The results of the radium benchmark dose calculations are used to establish a surface and subsurface soil dose limit for residual radionuclides other than radium, as well as a limit for surface activity on structures that will remain after decommissioning. The staff should review the licensee's conversion of the benchmark dose limit to soil concentration (pCi/g) or surface activity levels (dpm/100 cm^2) as a first step to determine cleanup levels. Alternatively, the licensee can derive the estimated dose from the uranium or thorium contamination (as discussed in Section 2.1.3) and compare this to the radium benchmark dose.

The reviewer should also evaluate the proposed cleanup guideline levels (derived concentration limit) in relation to the as low as is reasonably achievable requirement and the unity rule.

E2.2.2 Review Procedures

The decommissioning plan section on cleanup criteria will be evaluated for appropriate conversion of the radium standard benchmark dose to cleanup limits for soil uranium and thorium and/or surface activity. The plan will also be examined to ensure reasonable application of as low as is reasonably achievable to the cleanup guideline values and application of the unity rule where appropriate.

E2.2.3 Acceptance Criteria

(1)   The soil concentration limit is derived from the site radium dose estimate. The modeling performed to estimate mrem/year per pCi/g of Th-230 and/or U-nat follows the criteria listed in Section 2.1.3. In addition, the U-nat source term input is represented as percent activity by 48.9 percent U-238, 48.9 percent U-234, and 2.2 percent U-235,

or is based on analyses of the ore processed. For a soil uranium criterion (derived concentration limit), the chemical toxicity is considered in deriving a soil concentration limit if soluble forms of uranium are present.

(2)   Detailed justification for the inhalation pathway parameters is provided, such as the determination of the chemical form in the environment, to support the inhalation class.

(3)   The derived Th-230 soil limit will not cause any 100 square meter (m^2) area to exceed the Ra-226 limit at 1,000 years (i.e., current concentrations of Th-230 are less than 14 pCi/g surface and 43 pCi/g subsurface, if Ra-226 is at approximately background levels).

(4)   In conjunction with the activity limit, the as low as is reasonably achievable principle is considered in setting cleanup levels (derived concentration guideline levels). The as low as is reasonably achievable guidance in NUREG–1727, Appendix D (NRC, 2000) is considered. The proposed levels allow the licensee to demonstrate that the 10 CFR 40.42 (k) requirements (the premises are suitable for release, and reasonable effort has been made to eliminate residual radioactive contamination) can be met.

(5)   In recent practice at mill sites, the as low as is reasonably achievable principle is implemented by removing about 2 more inches [5 cm] of soil than is estimated to achieve the radium standard (reduce any possible excess or borderline contamination). At recovery facilities, it is generally cheaper to remove more soil than to do sampling and testing that may indicate failure and require additional soil removal with additional testing.

(6)   The unity rule is applied to the cleanup if more than one residual radionuclide is present in a soil verification grid (100 m^2). This means that the sum of the ratios for each radionuclide of the concentration present/concentration limit may not exceed 1 (i.e., unity).

(7)   The subsurface soil standard, if it is to be used, is applied to small areas of deep excavation where at least 15 cm [6 in.] of compacted clean fill is to be placed on the surface and where that depth of cover is expected to remain in place for the foreseeable future. The long-term cover depth used in the model is justified.

(8)   The surface activity limit for remaining structures is appropriately derived using an approved code or calculation. Because recent conservative dose modeling by NRC staff has indicated that more than 2,000 dpm/100 cm^2 alpha (U-nat or uranium chain radionuclides) in habitable buildings [2,000 hr/yr] could exceed an effective dose equivalent of 25 mrem/yr, the licensee proposes a total (fixed plus removable) average surface activity limit for such buildings that is lower than 2,000 dpm/100 cm^2, or a higher value is suitably justified.

(9)   If the DandD code is used, data are provided to support that 10 percent or less of the activity is removable; otherwise the resuspension factor is scaled to reflect the site-specific removable fraction. Note that this code assumes that the contamination is

only on the floor, which can be overly conservative. If the RESRAD-Build code is used, the modeled distribution of contamination on walls and floor is justified.

E2.2.4 Evaluation Findings

If the staff review, as described in this section, results in the acceptance of the application of the radium benchmark dose modeling to the site cleanup criteria, the following conclusions may be presented in the technical evaluation report.

The staff has completed its review of the proposed implementation of the benchmark dose modeling results for the _____ uranium *in situ* leach facility. This review included an evaluation using the review procedures and the acceptance criteria outlined in Section 2.2 of Appendix E of the *in situ* leach standard review plan.

The licensee has provided an acceptable implementation plan of the benchmark dose modeling results to the proposed site cleanup activities, and the staff evaluation determines that (1) the cleanup criteria will allow the licensee to meet 10 CFR Part 40.42(k) and 10 CFR Part 40, Appendix A, Criterion 6(6) requirements; (2) the soil and structures of the decommissioned site will permit termination of the license because public health and the environment will not be adversely affected by any residual radionuclides.

E3.0 <u>References</u>

Argonne National Laboratory. "Data Collection Handbook to Support Modeling the Impacts of Radioactive Material in Soil." ANL/EAIS–8. Washington, DC: U.S. Department of Energy. April 1993a.

———. "Manual for Implementing Residual Radioactive Material Guidelines Using RESRAD, Version 5.0." ANL/EAD/LD–2. Washington, DC: U.S. Department of Energy. September 1993b.

NRC. NUREG/CR–5512, "Residual Radioactive Contamination from Decommissioning—User's Manual." Volume 2. Washington, DC: NRC. April 2001.

———. NUREG–1727, "NMSS Decommissioning Standard Review Plan." Washington, DC: NRC. 2000.

———. Draft NUREG–1549, "Decision Methods for Dose Assessment to Comply With Radiological Criteria for License Termination." Washington, DC: NRC. July 1998.

———. NUREG/CR–5512, "Residual Radioactive Contamination from Decommissioning—Parameter Analysis." Draft. Volume 3. Washington, DC: NRC. April 1996.

Sandia National Laboratories. "Review of Parameter Data for the NUREG/CR–5512 Residential Farmer Scenario and Probability Distributions for the DandD Parameter Analysis." Draft Letter Report. January 30, 1998a.

———. "Review of Parameter Data for the NUREG/CR–5512 Building Occupancy Scenario and Probability Distributions for the DandD Parameter Analysis." Draft Letter Report. January 30, 1998b.

———. "Comparison of the Models and Assumptions Used in the DandD 1.0, RESRAD 5.61, and RESRAD-Build Computer Codes with Respect to the Residential Farmer and Industrial Occupant Scenarios Provided in NUREG/CR–5512." Draft Report. October 15, 1998c.